The Structure of Field Space

THE STRUCTURE OF FIELD SPACE

An Axiomatic Formulation of Field Physics

Dominic G. B. Edelen
The RAND Corporation

UNIVERSITY OF CALIFORNIA PRESS
Berkeley and Los Angeles 1962

University of California Press
Berkeley and Los Angeles, California
Cambridge University Press
London, England

Library of Congress Catalog Card Number: 62-16257
Printed in the United States of America

To Professor Albert L. Hammond
and the Proverbial Green Tree

Preface

THE MOTIVE BEHIND the various attempts to generalize the Einstein gravitational theory† is the prospect of obtaining a descriptive formalism that encompasses the dynamic interactions of material bodies, and that presents them as an orderly progression rather than as a set of intrinsically different effects. To achieve this goal, however, most of the attempts have been made in ways that might be described as putting the cart before the horse.

First, the attempt is to generalize a theory that offers adequate representations of the known facts concerning the little-understood phenomena of gravitation to the extent that it will offer representations of well-understood phenomena such as electromagnetism. If truly adequate representational formalisms are to be obtained, which correspond with the depth of our knowledge of physical processes, these formalisms must be required to admit the skew-symmetric and attendant structure necessary for representation of electromagnetic phenomena.

Second, gravitational and electromagnetic phenomena are structurally quite dissimilar. Gravitational forces can be eliminated at a point by an appropriate coördinate transformation, but electromagnetic forces cannot, in general, be so simply dismissed. The basic laws of electrodynamics may be displayed, without any assumptions of the metric properties of space-time, but the Einstein gravitational laws require specific metric structure; indeed, this metric structure is the seat of gravitational effects. The list of differences goes on and on.

Last, since the process of imbedding a given structure in a more elaborate one is anything but unique, there is at present an embarrassment of such imbeddings and, moreover, no clear-cut means of telling which is closest to the truth, with the exception of those that require the underlying space to admit intrinsic torsion. Spaces that admit intrinsic torsion are physically inadmissible since the invariant operation of coördinate-segment addition is noncommutative. If such noncommutativity were

† The major theories are listed in chronological order in the Bibliography at the end of the book.

allowed, the basis of our difference physics and the foundations of our theory of measurement would dissolve in confusion.

The line of attack adopted in the following pages is to examine the problem of field-theory construction from the other direction, namely, from the general to the particular.

Part One examines the structure of a class of field theories through an axiomatic approach. Since the class considered is quite large, even to the point of not requiring invariance structure, *the results obtained are of general validity and indicate just what can and what cannot be done mathematically without the use of specific physical assumptions.* Although some of the results are known, they have not previously been obtained without the assumption of invariance structure, nor have they been developed from an axiomatic formulation. The point of closest similarity is the affine field theory of Eddington, particularly in the identification of curvature with the Maxwellian field tensor.

The generality of the resulting "theory of field theories," which we refer to as the variant field theory, is just that—generality—unless it is shown to provide an adequate formalism in which the previous theories may be clothed.

Part Two demonstrates that variant field theory can be used to develop the classic field theories. This is demonstrated through explicitly stated constitutive assumptions, by which the relation of these theories with the general axiomatic structure may easily be perceived. It must be clearly noted that no claim to unification is made, and that no particular field theory is singled out as THE field theory. However, the gravitational, electromagnetic, and vector-meson field theories are obtained from a single Lagrangian function, so there are encouraging prospects for some degree of unification.

The following comments should be of use to those who wish quickly to acquaint themselves with the general content of this work. The exposition of the general theory is in Chapter III, and should be read in its entirety. Particular note should be taken of the fact that no invariance structure is required and that the class of Lagrangian functions is left quite general. Accordingly, this chapter may be used as an introduction to field-theoretic physics by the axiomatic method. The adequacy of the variant field theory as a descriptive formalism is demonstrated in Sections 4, 5, and 6 of Chapter VI and in Sections 3 and 4 of Chapter VII. Results established elsewhere are used in these sections. The reading of the cited sections may be accomplished with little difficulty, however, by tentatively accepting the validity of the required results. With more time available, the reader can then go back and examine the remainder of the text, thereby establishing for himself the validity of the tentatively accepted statements.

The author owes many thanks to a number of individuals who have been most generous with their time: To Dr. H. L. Leve and Dr. H. Bramson for their encouragement and discussions of many of the salient points, and to Dr. R. A. Toupin, Professor J. L. Ericksen, and Professor B. Hoffmann for their critical reviews and commentaries. The author is also indebted to the Hughes Aircraft Company for assistance during the initial phases of this work. The study was completed as part of the research program undertaken for the United States Air Force by The RAND Corporation.

D. G. B. E.

Santa Monica, Calif.
June 15, 1962

Contents

PART ONE

THE AXIOMATIC STRUCTURE OF FIELDS

SINCE THE publication of the general theory of relativity in 1916,† a number of theories have been proposed as generalizations of the Einstein theory or as alternatives designed to achieve a "unification" of the various domains of mathematical physics. With rare exceptions, these theories have resulted from attempts to imbed the equations of mathematical physics in suitably chosen geometric structures.

The approach we adopt in the construction of the "theory of variant fields" is to examine a class of possible field theories delineated by specifically stated axioms. We believe the axioms to be minimal, so that the variant field theory spans the universe of discourse open to possible field theories. Whether this is so or not, the class delineated by the employed axioms is extremely large and embraces most of the proposed theories. Thus variant field theory is not a particular field theory that makes any claim of unification; rather, it is a theory of the structure of field theories. The general results obtained in Chapters III and IV are therefore common to all field theories in the universe of discourse considered, and they show once and for all just which results are forthcoming in a natural mathematical way and which must be forced by restricting the universe of discourse through the addition of physical assumptions.

† A. Einstein: "Die Grundlage der allgemeinen Relativitätstheorie," *Ann. Physik*, **49** (1916), pp. 769–822.

Chapter I

Preliminary Considerations

MANY OF THE ideas of differential geometry, such as underlying topological or ground spaces, metric spaces, affinely connected spaces, infinitesimal parallel translation, and so forth, are familiar to the student of the various field theories, although not always in their axiomatic form. Since our purpose is to give an axiomatic foundation of field theories, we must first fix these basic geometric ideas in a precise manner.

1. Notation

Extensive use is made of tensor calculus throughout this book.† Greek covariant and contravariant indices are used to express tensors by means of their components with respect to an arbitrary coördinate system. Greek indices always follow the summation convention; that is, if the same index letter is used as a covariant and a contravariant index in a term, it is to be summed from 1 to 4.

Let $T_{\lambda\ldots\mu}$ be a set of terms with p indices $\lambda\ldots\mu$. Denote by $E_{\lambda\ldots\mu}$ $(O_{\lambda\ldots\mu})$ the sum of all T's with even (odd) permutations of the indices. The operations of *mixing* over p indices and *alternating* over p indices are defined by

$$p!T_{(\lambda\ldots\mu)} \overset{\text{def}}{=} E_{\lambda\ldots\mu} + O_{\lambda\ldots\mu}$$

and

$$p!T_{[\lambda\ldots\mu]} \overset{\text{def}}{=} E_{\lambda\ldots\mu} - O_{\lambda\ldots\mu},$$

respectively. Thus

† The reader unfamiliar with the general results of tensor calculus is referred to J. A. Schouten: *Ricci-Calculus, An Introduction to Tensor Analysis and Its Geometrical Applications*, Second Edition, Springer-Verlag, Berlin (1954); T. Y. Thomas: *The Differential Invariants of Generalized Spaces*, Cambridge University Press, London (1934).

$$6T_{(\lambda\eta\mu)} = (T_{\lambda\eta\mu} + T_{\eta\mu\lambda} + T_{\mu\lambda\eta}) + (T_{\eta\lambda\mu} + T_{\lambda\mu\eta} + T_{\mu\eta\lambda}),$$

$$6T_{[\lambda\eta\mu]} = (T_{\lambda\eta\mu} + T_{\eta\mu\lambda} + T_{\mu\lambda\eta}) - (T_{\eta\lambda\mu} + T_{\lambda\mu\eta} + T_{\mu\eta\lambda}).$$

We shall say that a tensor with components $T_{\alpha\beta\ldots\lambda}$ is symmetric (skew-symmetric) in the indices (α, β) if and only if

$$T_{[\alpha\beta]\ldots\lambda} = 0 \quad (T_{(\alpha\beta)\ldots\lambda} = 0).$$

It is to be noted that if the components $T_{\alpha\beta\gamma}$ are skew-symmetric in (α, β) then

$$3T_{[\alpha\beta\gamma]} = T_{\alpha\beta\gamma} + T_{\beta\gamma\alpha} + T_{\gamma\alpha\beta} = E_{\alpha\beta\gamma}.$$

To exclude certain indices from mixing or alternating, we enclose them by $|\quad|$; thus we have

$$2T_{(\alpha|\beta|\gamma)} = T_{\alpha\beta\gamma} + T_{\gamma\beta\alpha},$$

$$2T_{[\alpha|\beta|\gamma]} = T_{\alpha\beta\gamma} - T_{\gamma\beta\alpha}.$$

Although it is important to distinguish between a tensor and the set of elements that constitute its components, continual distinction quickly leads to an excessive use of the phrase "the components of." For this reason, we shall use the suffix notation to represent both the components of a tensor and the tensor itself. The context in which each interpretation of the suffix notation occurs should be adequate to distinguish between the two usages.

Partial differentiation will be denoted by a comma, and covariant differentiation formed from the affinity $L^{\alpha}{}_{\beta\gamma}$ will be denoted by a semicolon. When a partial derivative is to be taken with respect to an element of a collection of functions defined over the ground space (see following Sec. 2 for definition of ground space), it will be denoted by a comma followed by the symbol for the particular function as a subscript; thus we have

$$H(A, \mathbf{x})_{,A} = \partial H(A, \mathbf{x})/\partial A.$$

We adopt the following symbolic representations for volume and surface differentials:

$$dV \overset{\text{def}}{=} \prod_{\alpha=1}^{4} dx^{\alpha},$$

$$N_{\gamma}\, dS \overset{\text{def}}{=} \prod_{\alpha\neq\gamma}^{4} dx^{\alpha}.$$

Unless the contrary is explicitly stated, the radical sign $\sqrt{\ }$ or $(\ )^{1/2}$ always represents a nonnegative square root. To facilitate cross references, the following rule will be used: References not preceded by a chapter number refer to the chapter in which they occur.

2. Ground Spaces and Coördinates

The first notion we fix is that of an underlying or base space in which the normal processes of analysis are well defined.

DEFINITION 2.1. *By a base space $\mathcal{G}$ of n dimensions we shall mean a set of elements or points for which one can define a system of subsets called neighborhoods satisfying the following postulates:*

G_1: *The points of each neighborhood can be put into a bijective (one-one) correspondence with the interior points of a hypersphere of n-dimensional Euclidean space.*

G_2: *Each point of $\mathcal{G}$ belongs to at least one neighborhood.*

G_3: *For any two neighborhoods of any given point of $\mathcal{G}$ there is a neighborhood contained in the intersection of the two given neighborhoods.*

G_4: *If a point b is contained in a neighborhood $N(a)$ of a point a, then there is a neighborhood $N(b)$ of b contained in $N(a)$.*

G_5: *If a and b are any two distinct points of $\mathcal{G}$, then there are neighborhoods $N(a)$ and $N(b)$, of a and b, respectively, such that $N(a)$ and $N(b)$ have no points in common.*†

Postulates G_2, G_3, and G_4 state that $\mathcal{G}$ is a topological space with basic sets defined by postulate G_1. Postulate G_5 establishes the Hausdorff structure of $\mathcal{G}$.

A point a of $\mathcal{G}$ is said to be a *cluster point* for an infinite set of points of $\mathcal{G}$ if each neighborhood containing a contains at least one point of the set distinct from a. We shall accordingly say that the space $\mathcal{G}$ is *open*, or *closed*, if it has, or does not have, an infinite set of points admitting no cluster point.

An infinite sequence of points $a_1, a_2, \ldots, a_n, \ldots$ is said to approach a *limit a* if, given any arbitrary neighborhood $N(a)$, there is a point a_n such that all following points of the sequence lie in $N(a)$. It follows from G_5 that an infinite sequence can have at most one limit point.

By a continuous curve is meant a set of points that can be put into a

† This set of postulates is essentially that used by E. Cartan: "La Théorie des Groupes Finis et Continus et l'Analysis Situs," *Mémor. Sci. Math.*, No. 24, Gauthier-Villars et Cie., Paris (1930).

bijective correspondence with the numerical values of a real variable t on the interval $0 \leqq t \leqq 1$, such that if a sequence t_n has a limit point t_0, the sequence of points corresponding to t_n has a limit point corresponding to t_0. A space $\mathcal{G}$ is said to be *connected* if any two arbitrary points of $\mathcal{G}$ can be joined by a continuous curve, all of whose points are in $\mathcal{G}$.

We shall now define the ground space that will be used throughout this book.

DEFINITION 2.2. *By a ground space $\mathcal{E}$ we shall mean a space satisfying the following postulates:*

E_1: *$\mathcal{E}$ is a 4-dimensional $\mathcal{G}$.*
E_2: *$\mathcal{E}$ is connected.*

By postulates G_1, G_2, G_3, G_4, and E_1, a bijective correspondence exists between the points of a neighborhood of $\mathcal{E}$ and the interior points of a 4-dimensional hypersphere of Euclidean 4-space. Hence the points of a neighborhood can be defined by 4 real coördinates $x^1, \ldots, x^4$ such that if a is the limit point of a sequence a_n lying in the neighborhood, the distance from a to a_n, as given by the Euclidean measure of distance, approaches zero as n approaches infinity. Any particular association of the coördinates $x^1, \ldots, x^4$ with the points of a neighborhood is called a *coördinate system*, the (x) coördinate system.

The process of passing from one coördinate system to another is called a *transformation of coördinates*. Such a transformation is represented by a set of four equations

$$x^\alpha = f^\alpha('x^1, \ldots, 'x^4), \quad \alpha = 1, \ldots, 4, \tag{2.1}$$

which state that the coördinates x^α of a point a with respect to the (x) system of coördinates are determined when the coördinates $'x^\alpha$ of the same point a with respect to the $('x)$ coördinate system are given. Coördinate transformations will be said to be *admissible* if and only if the functions $f^\alpha('\mathbf{x})$ possess continuous derivatives of order p ($p \geqq 1$) and the Jacobian matrix

$$((j^\alpha_{'\beta})) \overset{\text{def}}{=} \left(\left(\frac{\partial x^\alpha}{\partial' x^\beta}\right)\right) \tag{2.2}$$

is nonsingular. Only admissible coördinate transformations will be considered. Under the requirement of admissibility, (2.1) can be inverted, at least locally, to yield

$$'x^\alpha = F^\alpha(x^1, \ldots, x^4), \quad \alpha = 1, \ldots, 4. \tag{2.3}$$

Similarly, there exists a unique inverse Jacobian matrix

$$((j'^{\alpha}_{\beta})) \overset{\text{def}}{=} \left(\left(\frac{\partial' x^{\alpha}}{\partial x^{\beta}}\right)\right) \tag{2.4}$$

with the properties

$$j'^{\alpha}_{\beta} j^{\beta}_{\cdot\gamma} = \delta^{\alpha}_{\gamma}, \qquad j'^{\alpha}_{\beta} j^{\gamma}_{\cdot\alpha} = \delta^{\gamma}_{\beta}. \tag{2.5}$$

We shall now show that any compact set of $\mathcal{E}$ can be covered by a finite number of coördinate systems, related by analytic transformations in regions where they overlap, and thus obtain the normal coördinate representations of differential geometry. Let $\mathcal{H}$ be any compact set of $\mathcal{E}$. By postulates E_1 and G_2, every point of $\mathcal{H}$ lies in at least one neighborhood. We can thus assign at least one neighborhood to each point of $\mathcal{H}$ and thereby cover $\mathcal{H}$ by a collection of open sets such that each point is interior to at least one element of this collection. (Neighborhoods are open sets since they are in bijective correspondence with the *interior* points of Euclidean 4-spheres.) Since $\mathcal{H}$ is assumed to be compact, the Heine–Borel theorem is applicable, and hence there exists a finite collection of neighborhoods covering $\mathcal{H}$. Each neighborhood of this finite collection can be used to assign a coördinate system to its contained points, as above, and hence we have a finite number of coördinate systems over $\mathcal{H}$. Since each of these coördinate systems is in bijective correspondence with the coördinate system of a Euclidean 4-sphere, they are equivalent, and hence can be so chosen that there exist analytic coördinate transformations between any two of the finite collection of coördinate systems in regions where they overlap.

A point set in $\mathcal{E}$ will be referred to as a *domain* D if it is an open, connected, nonempty point set with compact closure D^*. Since D^* is a connected compact set, it can be covered by a finite number of coördinate systems. We shall always assume that this has been done. By the *boundary* of D will be meant the set $D^* \ominus D$, where $\ominus$ stands for the set-theoretic difference.

3. Affinely Connected Spaces

A collection of spaces having central importance in the following chapters consists of those spaces that are affinely connected.

Definition 3.1. *A ground space is said to be affinely connected if and only if there is defined at each point a collection of 64 functions* $A^{\alpha}{}_{\beta\gamma}(\mathbf{x})$, *called the components of affine connection, which transform according to the law*

$$'A^{\alpha}{}_{\beta\gamma} = j'^{\alpha}_{\eta}\left(A^{\eta}_{\mu\nu}\, j'^{\mu}_{\beta}\, j'^{\nu}_{\gamma} + \frac{\partial^2 x^{\eta}}{\partial' x^{\beta}\, \partial' x^{\gamma}}\right) \tag{3.1}$$

under admissible coördinate transformations $'x^{\alpha} = f^{\alpha}(x^{\beta})$.

As we know, the property of affine connectivity of a space allows us to speak of neighboring vectors as parallel, to perform the operation of covariant differentiation, and to construct the curvature tensors that we shall discuss in detail in the next chapter. (A vector $T^{\alpha}(\mathbf{x})$ is said to be parallel to a vector $T^{\alpha}(\mathbf{x} + d\mathbf{x})$ if $T^{\alpha}(\mathbf{x} + d\mathbf{x}) = T^{\alpha}(\mathbf{x}) - A^{\alpha}{}_{\beta\gamma}(\mathbf{x})\, T^{\beta}(\mathbf{x})\, dx^{\gamma}$.)

The components of affine connection $A^{\alpha}{}_{\beta\gamma}$ can be broken into two parts by putting

$$A^{\alpha}{}_{\beta\gamma} = L^{\alpha}{}_{\beta\gamma} + X^{\alpha}{}_{\beta\gamma}, \tag{3.2}$$

where

$$L^{\alpha}{}_{\beta\gamma} \overset{\text{def}}{=} A^{\alpha}{}_{(\beta\gamma)} \tag{3.3}$$

and

$$X^{\alpha}{}_{\beta\gamma} \overset{\text{def}}{=} A^{\alpha}{}_{[\beta\gamma]}. \tag{3.4}$$

If the quantities $X^{\alpha}{}_{\beta\gamma}$ vanish, the affine connection is said to be *symmetric;* otherwise it is said to be *asymmetric.* It is to be noted that $X^{\alpha}{}_{\beta\gamma}$ is a tensor, since the last term in (3.1) vanishes under alternation over (β, γ); it is referred to as the *torsion tensor.*

We shall be concerned with those affinely connected spaces $\mathcal{A}$ for which the torsion tensor vanishes, in contrast to the spaces of the Einstein–Hlavatý–Schrödinger theory.† The bases for rejecting spaces with intrinsic torsion are: First, we are able to obtain all the required results and structure for adequate representations of both gravitational and electromagnetic phenomena, without the additional complexities associated with spaces of nonvanishing torsion. Second, spaces with intrinsic torsion seem physically inadmissible since the operation of invariant geometric addition of infinitesimal coördinate segments is noncommutative, a result in direct conflict with the operational and epistemological basis of our difference physics.

† V. Hlavatý: *Geometry of Einstein's Unified Field Theory*, P. Noordhoff, Ltd., Groningen, Holland (1957).

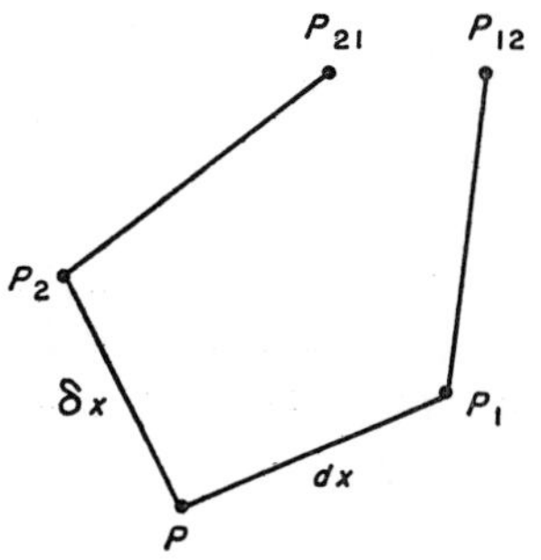

To demonstrate this noncommutativity, let P be an arbitrary point of an affinely connected space with coördinates x^α, and consider two arbitrary infinitesimal vectors d and δ at P with components dx^α and δx^α, respectively (see illustration). These two vectors define two points P_1 and P_2 in a neighborhood of P with coördinates $x^\alpha + dx^\alpha$ and $x^\alpha + \delta x^\alpha$, respectively, and hence dx^α and δx^α may be interpreted as infinitesimal coördinate segments. Since the space under consideration is assumed to be affinely connected, we can define the addition of the coördinate segments dx^α and δx^α in an invariant geometric manner:

$$dx^\alpha \oplus \delta x^\alpha \overset{\text{def}}{=} dx^\alpha + (\text{the } \alpha\text{-th component of } \delta x \text{ obtained after parallel translation of } \delta x \text{ along } dx \text{ to the extremity of } dx).$$

(Although it is not usually recognized, this is the definition of addition that is most often used in the laboratory in assigning coördinate values to physical events.) Under this definition, we have

$$dx^\alpha \oplus \delta x^\alpha = dx^\alpha + \delta x^\alpha - A^\alpha{}_{\beta\gamma}\, \delta x^\beta\, dx^\gamma$$

on neglecting higher-order terms, so that the coördinates of the point P_{12} at the extremity of $dx^\alpha \oplus \delta x^\alpha$ are given by $x^\alpha + dx^\alpha + \delta x^\alpha - A^\alpha{}_{\beta\gamma}\, \delta x^\beta\, dx^\gamma$. On the other hand, we have

$$\delta x^\alpha \oplus dx^\alpha = \delta x^\alpha + dx^\alpha - A^\alpha{}_{\beta\gamma}\, dx^\beta\, \delta x^\gamma$$

on neglecting higher-order terms, so that the coördinates of the point P_{21} at the extremity of $\delta x^\alpha \oplus dx^\alpha$ are given by $x^\alpha + dx^\alpha + \delta x^\alpha - A^\alpha{}_{\beta\gamma}\, dx^\beta\, \delta x^\gamma$. The difference in the coördinates of P_{12} and P_{21} is thus directly given in terms of the commutator $<d| \oplus |\delta>^\alpha$ of d and δ with respect to the operation $\oplus$, as defined by

$$<d| \oplus |\delta>^\alpha \overset{\text{def}}{=} (dx^\alpha \oplus \delta x^\alpha) - (\delta x^\alpha \oplus dx^\alpha). \tag{3.5}$$

Substituting from the above relations gives

$$\begin{aligned} <d| \oplus |\delta>^{\alpha} &= -A^{\alpha}{}_{\beta\gamma}\, \delta x^{\beta}\, dx^{\gamma} + A^{\alpha}{}_{\beta\gamma}\, dx^{\beta}\, \delta x^{\gamma} \\ &= (L^{\alpha}{}_{\beta\gamma} - L^{\alpha}{}_{\gamma\beta})\, dx^{\beta}\, \delta x^{\gamma} + (X^{\alpha}{}_{\beta\gamma} - X^{\alpha}{}_{\gamma\beta})\, dx^{\beta}\, \delta x^{\gamma} \\ &= 2X^{\alpha}{}_{\beta\gamma}\, dx^{\beta}\, \delta x^{\gamma} \end{aligned}$$

by use of (3.2)–(3.4). Thus, since the vectors d and δ are arbitrary to within the requirement that they be infinitesimal, P_{12} and P_{21} are the same point, and invariant geometric addition is commutative if and only if the torsion tensor $X^{\alpha}{}_{\beta\gamma}$ vanishes identically.

4. Metric Spaces

One of the essential features of the variant field theory presented in Chapter III is that it makes no assumptions concerning the metric structure of space. In fact, if the ground space were assumed to be metric, one of the essential building tensors of the theory would be lost. (See III-3.) In order to avoid any doubt of what we mean by a metric space, we give the following definition.

DEFINITION 4.1. *An affinely connected space is a metric space* $\mathfrak{M}$ *if and only if there is defined in every D a tensor* $g_{\alpha\beta}(x)$ *satisfying the following postulates:*

M_1: $g_{\alpha\beta}$ *is symmetric, that is,* $g_{[\alpha\beta]} = 0$.
M_2: $g \overset{\text{def}}{=} \det(g_{\alpha\beta}) \neq 0$.
M_3: $g_{\alpha\beta;\gamma} = 0$.

These postulates are equivalent to the usual differential geometric definition of a metric space, in that length, vector magnitude, and angle are well-defined quantities. (Since $g_{\alpha\beta;\gamma} = 0$, it follows that

$$ds^2 = g_{\alpha\beta}\, dx^{\alpha}\, dx^{\beta}, \quad |F| = g_{\alpha\beta}\, F^{\alpha}\, F^{\beta},$$

and

$$\cos\theta = g_{\alpha\beta}\, F^{\alpha}\, T^{\beta}\, |F|^{-1/2}\, |T|^{-1/2}$$

are well-defined geometric quantities with the properties normally associated with distance, vector magnitude, and angle determination provided neither $|F|$ nor $|T|$ vanishes.†) It must be noted that they are not

† T. Y. Thomas: *op. cit.*, pp. 10–15.

equivalent to the postulates of a metric space in the sense of a metric topological space, since it is not assumed that $g_{\alpha\beta}$ is positive definite.

Postulate M_3 states that the affine connection of an $\mathfrak{M}$ is determined uniquely by $g_{\alpha\beta}$. To see this, we have only to expand

$$g_{\alpha\beta;\gamma} = 0$$

to obtain

$$g_{\alpha\beta,\gamma} - A^{\rho}{}_{\gamma\alpha}\, g_{\rho\beta} - A^{\rho}{}_{\gamma\beta}\, g_{\alpha\rho} = 0$$

and then apply Christoffel's eliminative method.† This gives

$$A^{\alpha}{}_{\beta\gamma} = \tfrac{1}{2}\, g^{\alpha\rho}\, (g_{\rho\beta,\gamma} + g_{\rho\gamma,\beta} - g_{\beta\gamma,\rho}),$$

so that the $A^{\alpha}{}_{\beta\gamma}$, which are obviously symmetric, are the Christoffel symbols of the second kind formed from $g_{\alpha\beta}$. In applying Christoffel's eliminative method, we have of course used postulates M_1 and M_2 to obtain the unique tensor $g^{\alpha\beta}$ reciprocal to $g_{\alpha\beta}$ given by $g\, g^{\alpha\beta} = g_{,g_{\alpha\beta}}$.

5. Well-based Spaces

A variety of spaces, in some ways similar to metric spaces, will arise in many of our considerations.

DEFINITION 5.1. *An affinely connected space will be said to be a well-based space* $\mathfrak{B}$ *with base field* $h_{\alpha\beta}$ *if and only if the following postulates are satisfied:*

B_1: *There is defined in each* $D \subset \mathfrak{B}$ *a symmetric tensor* $h_{\alpha\beta}$; *that is,* $h_{\alpha\beta}$ *satisfies the equation* $h_{[\alpha\beta]} = 0.$ (5.1)

B_2: $h \overset{\text{def}}{=} \det(h_{\alpha\beta}) \neq 0$ *everywhere in* $\mathfrak{B}$. (5.2)

B_3: $h_{\alpha\beta;\gamma} = -Q_{\gamma\alpha\beta}$, *where the* $Q_{\gamma\alpha\beta}$ *are of class* C^1 *in* $\mathfrak{B}$. (5.3)

Under these postulates, it can be shown that a symmetric affine connection $L^{\alpha}{}_{\beta\gamma}$ is uniquely determined as a function of $h_{\alpha\beta}$ and $Q_{\gamma\alpha\beta}$ and is given by‡

$$L^{\alpha}{}_{\beta\gamma} = \Gamma^{\alpha}{}_{\beta\gamma}(h_{\eta\pi}) + \tfrac{1}{2}\, h^{\alpha\rho}\, (Q_{\beta\rho\gamma} + Q_{\gamma\beta\rho} - Q_{\rho\gamma\beta}), \tag{5.4}$$

† See P. G. Bergmann: *Introduction to the Theory of Relativity*, Prentice-Hall, Inc., Englewood Cliffs, N. J. (1942), p. 72.

‡ J. A. Schouten: *op. cit.*, p. 132.

where the $\Gamma^\alpha{}_{\beta\gamma}(h_{\eta\pi})$ are the Christoffel symbols of the second kind based on the $h_{\alpha\beta}$ field. To prove that (5.4) defines a symmetric affine connection, we need only to compute the torsion tensor $L^\alpha{}_{[\beta\gamma]}$ from (5.4) and note that $Q_{\gamma[\alpha\beta]} = 0$ by postulates B_1 and B_3 to see that $L^\alpha{}_{[\beta\gamma]}$ vanishes.

Chapter II

Curvature and Related Topics

THE NOTION of the curvature of a curve is a familiar one that is encountered in many subject areas such as the arc-parameter kinematics of particle motion and the Bernoulli–Euler theory of beam deformation. Similarly, the notion of the curvature of a 2-surface in a 3-dimensional Euclidean space is basic in such everyday problems as the analysis of the bending of plate and shell structures. In differential geometry and geometric field theory, the concept of curvature of the ground space is a basic notion, fundamental in both the characterization of the ground space in terms of its intrinsic structure and in the characterization of geometric field theories. Since the general spaces we shall be concerned with are affinely connected, but not necessarily metric, many of the formulas and results familiar to the student of general relativity are not valid here. This is particularly true for curvature expressions. Let us examine curvature expressions in a general space of symmetric affine connection to lay the appropriate ground work for succeeding considerations.

1. The Curvature Tensor

Let the underlying space be an $\mathcal{E}$ (a space satisfying the postulates E_1 and E_2 of I-2) into which a structure is introduced by requiring that a *symmetric* affine connection $L^{\alpha}{}_{\beta\gamma}$ be defined at each point of $\mathcal{E}$. The quantity $K^{\eta}{}_{\alpha\beta\gamma}$, defined by

$$K^{\eta}{}_{\alpha\beta\gamma} \stackrel{\text{def}}{=} 2L^{\eta}{}_{\alpha[\gamma,\beta]} + 2L^{\eta}{}_{\mu[\beta}\, L^{\mu}{}_{\gamma]\alpha}, \tag{1.1}$$

is the affine curvature tensor of ground spaces $\mathcal{A}$ with symmetric affine connection $L^{\alpha}{}_{\beta\gamma}$. To see that $K^{\eta}{}_{\alpha\beta\gamma}$ is in actuality a tensor, note that by the definition of the covariant derivative of an arbitrary vector N^{λ} we have

$$2N^{\lambda}{}_{;[\beta;\gamma]} = -K^{\lambda}{}_{\alpha\beta\gamma}\, N^{\alpha}. \tag{1.2}$$

Thus, applying the quotient theorem and noting that N^α and $N^\lambda{}_{;[\beta;\gamma]}$ are tensors, we immediately establish the tensorial character of $K^\eta{}_{\alpha\beta\gamma}$.

Under the assumptions defining $\mathfrak{A}$, the curvature tensor given by (1.1) satisfies the algebraic identities

$$K^\alpha{}_{\beta(\gamma\lambda)} = 0, \qquad K^\alpha{}_{[\beta\gamma\lambda]} = 0, \tag{1.3}$$

which are a complete set.† If the ground space has a more constrained structure than that of $\mathfrak{A}$, the number of identities constituting a complete set is in general increased. For instance, in the complete set for a metric space, there are two identities‡ in addition to those given in (1.3).

In later sections, we shall see that many spaces of interest are well-based spaces $\mathfrak{B}$ (spaces satisfying postulates B_1, B_2, B_3 of I-2). Let $h_{\alpha\beta}$ and $Q_{\gamma\alpha\beta}$ be as defined by (I-5.1)–(I-5.3) and set

$$T_{\mu\lambda}^{\cdot\cdot\alpha} \stackrel{\text{def}}{=} \tfrac{1}{2}\, h^{\rho\alpha}\, Q_{\{\mu\rho\lambda\}}, \tag{1.4}$$

$$\psi_{\{\alpha\beta\gamma\}} \stackrel{\text{def}}{=} \psi_{\alpha\beta\gamma} - \psi_{\beta\gamma\alpha} + \psi_{\gamma\alpha\beta}. \tag{1.5}$$

Denote by $\Gamma^\alpha{}_{\beta\gamma}$ the Christoffel symbols formed from $h_{\alpha\beta}$, namely

$$\Gamma^\alpha{}_{\beta\gamma} \stackrel{\text{def}}{=} \tfrac{1}{2}\, h^{\rho\alpha}\,(h_{\rho\beta,\gamma} + h_{\rho\gamma,\beta} - h_{\beta\gamma,\rho}), \tag{1.6}$$

and let $J^\eta{}_{\alpha\beta\gamma}$ represent the curvature tensor formed from $\Gamma^\alpha{}_{\beta\gamma}$. Then we have

$$J^\eta{}_{\alpha\beta\gamma} \stackrel{\text{def}}{=} 2\Gamma^\eta{}_{\alpha[\gamma,\beta]} + 2\Gamma^\eta{}_{\mu[\beta}\,\Gamma^\mu{}_{\gamma]\alpha}$$

$$(J_{\alpha\beta} = J^\eta{}_{\alpha\beta\eta}, \quad J^\eta{}_{\eta\beta\gamma} = 0), \tag{1.7}$$

which is the curvature tensor of a metric space $\mathfrak{M}$ with metric tensor $h_{\alpha\beta}$. It can then be shown§ that for a well-based space we have

$$K^\alpha{}_{\lambda\nu\mu} = J^\alpha{}_{\lambda\nu\mu} + 2\nabla_{[\nu} T_{\mu]\lambda}^{\cdot\cdot\alpha} + 2T_{[\nu|\rho|}^{\cdot\cdot\alpha}\, T_{\mu]\lambda}^{\cdot\cdot\rho} \tag{1.8}$$

where ∇_ν stands for the covariant derivative formed from the affinity $\Gamma^\alpha{}_{\beta\gamma}$, and the dots stand for indices that have been pulled by $h_{\alpha\beta}$ or $h^{\alpha\beta}$. Equation (1.8) will be of considerable use in later sections. It will allow us to interpret many of the results of variant field theory in terms of

† T. Y. Thomas: *op. cit.*, p. 132. A system of algebraic identities is said to be complete if and only if there exists no algebraic identity that cannot be represented as an algebraic function of the system of identities considered.

‡ J. A. Schouten: *op. cit.*, pp. 144–153.

§ *Ibid.*, p. 141.

operations in a subsidiary 4-dimensional metric space of the Riemannian variety and thereby derive the metric properties that are held by many to be required of a physical theory.

Let $\delta L^{\alpha}{}_{\beta\gamma}$ be an arbitrary tensor of class C^1 that induces a variation of the affine connection $L^{\alpha}{}_{\beta\gamma}$ and preserves the symmetry of that connection, so that

$$\delta L^{\alpha}{}_{[\beta\gamma]} = 0.$$

(That the variation of an affine connection is always a tensor is seen as follows. We have

$$*L^{\alpha}{}_{\beta\gamma} = L^{\alpha}{}_{\beta\gamma} + \delta L^{\alpha}{}_{\beta\gamma},$$

so that

$$\delta L^{\alpha}{}_{\beta\gamma} = *L^{\alpha}{}_{\beta\gamma} - L^{\alpha}{}_{\beta\gamma},$$

where both $*L^{\alpha}{}_{\beta\gamma}$ and $L^{\alpha}{}_{\beta\gamma}$ transform according to (I-3.1). Thus, under a coördinate transformation, we have

$$'\delta L^{\alpha}{}_{\beta\gamma} = j^{\prime\alpha}_{\eta}\left[(*L^{\eta}{}_{\mu\nu} - L^{\eta}{}_{\mu\nu})\, j^{\mu}_{\prime\beta} j^{\nu}_{\prime\gamma} + \frac{\partial^2 x^{\eta}}{\partial' x^{\beta}\, \partial' x^{\gamma}} - \frac{\partial^2 x^{\eta}}{\partial' x^{\beta}\, \partial' x^{\gamma}}\right]$$

$$= j^{\prime\alpha}_{\eta} j^{\mu}_{\prime\beta} j^{\nu}_{\prime\gamma}\, \delta L^{\eta}_{\mu\nu},$$

as desired.) By use of equation (1.1), we can easily show through direct calculation that the variation $\delta K^{\alpha}{}_{\beta\gamma\lambda}$ induced in the affine curvature tensor by the variation $\delta L^{\alpha}{}_{\beta\gamma}$ of the affine connection is given by

$$\delta K^{\alpha}{}_{\beta\gamma\lambda} = (\delta L^{\alpha}{}_{\beta\lambda})_{;\gamma} - (\delta L^{\alpha}{}_{\beta\gamma})_{;\lambda}. \tag{1.9}$$

2. Contractions of the Curvature Tensor and Space Structure

We can build three new tensors in $\mathfrak{A}$ from the tensor $K^{\alpha}{}_{\beta\gamma\lambda}$ by contraction, as follows:

$$A_{\alpha\beta} \stackrel{\text{def}}{=} K^{\eta}{}_{\eta\alpha\beta} = 2L^{\eta}{}_{\eta[\beta,\alpha]}, \tag{2.1}$$

$$B_{\alpha\beta} \stackrel{\text{def}}{=} K^{\eta}{}_{\alpha\eta\beta} = 2L^{\eta}{}_{\alpha[\beta,\eta]} + 2L^{\eta}{}_{\mu[\eta}\, L^{\mu}{}_{\beta]\alpha}, \tag{2.2}$$

$$C_{\alpha\beta} \stackrel{\text{def}}{=} K^{\eta}{}_{\alpha\beta\eta} = 2L^{\eta}{}_{\alpha[\eta,\beta]} + 2L^{\eta}{}_{\mu[\beta}\, L^{\mu}{}_{\eta]\alpha}. \tag{2.3}$$

These three tensors are not algebraically independent, however, because of the algebraic identities (1.3) satisfied by $K^{\alpha}{}_{\beta\gamma\lambda}$. Substituting (2.1)–(2.3) into (1.3), we have

$$B_{\alpha\beta} = -C_{\alpha\beta}, \qquad A_{\alpha\beta} = -2C_{[\alpha\beta]}. \tag{2.4}$$

Thus there is only one algebraically independent contraction, say, $C_{\alpha\beta}$, of the affine curvature tensor, the other two being obtained from it by (2.4).

Rather than work with the tensor $C_{\alpha\beta}$ directly, we decompose it into its skew-symmetric and symmetric parts, as defined by

$$A_{\alpha\beta} = -2C_{[\alpha\beta]} = L^{\eta}{}_{\eta[\beta,\alpha]}, \qquad A_{(\alpha\beta)} = 0, \tag{2.5}$$

and

$$R_{\alpha\beta} \overset{\text{def}}{=} C_{(\alpha\beta)} = L^{\eta}{}_{\eta(\alpha,\beta)} - L^{\eta}{}_{\alpha\beta,\eta} + 2L^{\rho}{}_{\eta[\beta}\, L^{\eta}{}_{\rho]\alpha}, \quad R_{[\alpha\beta]} = 0, \tag{2.6}$$

respectively, and work with these tensors separately. The above decomposition of $C_{\alpha\beta}$ into an independent symmetric tensor $R_{\alpha\beta}$ and an independent skew-symmetric tensor $A_{\alpha\beta}$ will allow us to examine the properties of the ground space $\mathfrak{A}$ that are particularized by the property of symmetry or skew-symmetry. This decomposition of $C_{\alpha\beta}$ also admits a simple and natural separation of electromagnetic effects and nonelectromagnetic effects in the field theory that is discussed in succeeding chapters of this book.

The vanishing or nonvanishing of the tensor $A_{\alpha\beta}$ determines certain intrinsic properties of the space $\mathfrak{A}$, as the following results show.

THEOREM 2.1. *A necessary condition for the existence of a symmetric, nonsingular, minimally constrained covariant constant tensor field $T_{\alpha\beta}$ of weight $w \neq -1/2$ in $\mathfrak{A}$ is $A_{\alpha\beta} = 0$.*

Proof. By definition, a tensor field $T_{\alpha\beta}$ of weight w forms a minimally constrained covariant constant field in $\mathfrak{A}$ if and only if the equations

$$T_{\alpha\beta;\delta} = 0 \tag{2.7}$$

are completely integrable. A necessary and sufficient condition that the system (2.7) be completely integrable is given by

$$T_{\alpha\beta;[\delta\gamma]} = 0. \tag{2.8}$$

Expanding the left-hand side of (2.8) by means of the definition of the covariant derivative and using (1.1), we have†

$$T_{\alpha\beta;[\delta\gamma]} = T_{\sigma(\alpha} K^{\sigma}{}_{\beta)\delta\gamma} + 1/2\, w\, T_{\alpha\beta} A_{\delta\gamma};$$

hence a necessary and sufficient condition for the existence of a minimally constrained covariant constant field $T_{\alpha\beta}$ in $\mathfrak{A}$ is

$$T_{\sigma(\alpha} K^{\sigma}{}_{\beta)\delta\gamma} + 1/2\, w\, T_{\alpha\beta} A_{\delta\gamma} = 0. \tag{2.9}$$

Since $T_{\alpha\beta}$ is nonsingular, by hypothesis, there exists a unique field $T^{\alpha\beta}$ with the property

$$T_{\alpha\beta} T^{\beta\gamma} = \delta^{\gamma}_{\alpha}.$$

Multiplication of (2.9) by $T^{\alpha\beta}$ gives

$$K^{\sigma}{}_{\sigma\delta\gamma} + 2w\, A_{\delta\gamma} = 0.$$

By (2.1), this implies that

$$(1 + 2w)\, A_{\alpha\beta} = 0,$$

and hence the theorem is established.

Although the following result is really a corollary of Theorem 2.1, it is stated as a theorem because of its importance in later chapters.

THEOREM 2.2. *A necessary condition for the reduction of an affinely connected space* $\mathfrak{A}$ *to a metric space* $\mathfrak{M}$ *is* $A_{\alpha\beta} = 0$.

Proof. The only structure assumed for the affinely connected space $\mathfrak{A}$ is the affine connection $L^{\alpha}{}_{\beta\gamma}$, while for a metric space the existence of a tensor field $g_{\alpha\beta}$ satisfying postulates M_1, M_2, M_3 is assumed. We therefore have to examine the conditions that must be placed on the $L^{\alpha}{}_{\beta\gamma}$ of the space $\mathfrak{A}$ in order that the existence of a tensor $g_{\alpha\beta}$ with these properties is implied. Let $h_{\alpha\beta}$ be a symmetric, nonsingular tensor field of weight zero. Such a field satisfies postulates M_1 and M_2. Postulate M_3 requires

$$h_{\alpha\beta;\gamma} = 0$$

if $h_{\alpha\beta}$ is to be identified with the metric tensor $g_{\alpha\beta}$. Since a metric tensor is

† J. A. Schouten: *op. cit.*, p. 140. It should be noted that Schouten's $R_{\alpha\beta\gamma}{}^{\mu}$ is equal to our $K^{\mu}{}_{\gamma\alpha\beta}$, his $R_{\alpha\beta}$ to the negative of our $R_{\alpha\beta}$, and his $V_{\alpha\beta}$ to our $A_{\alpha\beta}$.

also minimally constrained, the result follows as a direct consequence of Theorem 2.1.

THEOREM 2.3. *A necessary and sufficient condition for the existence of a minimally constrained covariant constant scalar density field in* $\mathfrak{A}$ *is* $A_{\alpha\beta} = 0$.†

Proof. By definition, a scalar density $D(\mathbf{x})$ forms a minimally constrained covariant constant field in $\mathfrak{A}$ if and only if

$$D_{;\gamma} \stackrel{\text{def}}{=} D_{,\gamma} - L^{\eta}{}_{\eta\gamma}\, D = 0$$

is completely integrable. This equation is completely integrable for $\ln(D)$, and hence for D, if and only if

$$L^{\eta}{}_{\eta[\gamma,\lambda]} = 0,$$

which implies $A_{\alpha\beta} = 0$ by (2.5).

In this paragraph, we collect certain results that will be required later. The forms that $A_{\alpha\beta}$ and $R_{\alpha\beta}$ take in a well-based space $\mathfrak{B}$ can be obtained directly from (2.1), (2.3), (2.6), and (1.8):‡

$$A_{\tau\mu} = 2(h^{\rho\xi}\, T_{[\mu|\xi\rho|})_{,\tau]}, \tag{2.10}$$

$$R_{\mu\lambda} = J_{\mu\lambda} - h^{\rho\xi}\, \nabla_{\xi} T_{\dot{\mu}\dot{\lambda}\dot{\rho}} + h^{\rho\xi}\, \nabla_{(\mu} T_{|\dot{\xi}|\dot{\lambda})\dot{\rho}}$$
$$- 2h^{\rho\xi}\, h^{\epsilon\tau}\, T_{[\dot{\tau}|\dot{\rho}\dot{\epsilon}|}\, T_{\dot{\mu}]\dot{\lambda}\dot{\xi}}. \tag{2.11}$$

The variations induced in $A_{\alpha\beta}$ and $R_{\alpha\beta}$ by a variation of $L^{\alpha}{}_{\beta\gamma}$ may be obtained directly from (2.1), (2.3), (2.6), and (1.9):

$$\delta A_{\alpha\beta} = 2(\delta L^{\eta}{}_{\eta[\beta})_{;\alpha]}, \tag{2.12}$$

$$\delta R_{\alpha\beta} = (\delta L^{\eta}{}_{\eta(\alpha})_{;\beta)} - (\delta L^{\eta}{}_{\alpha\beta})_{;\eta}. \tag{2.13}$$

3. Differential Identities

In addition to the algebraic identities of the curvature tensor of $\mathfrak{A}$, there is a system of differential identities that is of considerable importance. From (1.1) we have the equality

† J. A. Schouten: *op. cit.*, p. 155.
‡ *Ibid.*, p. 141.

$$L^{\alpha}{}_{\lambda[\mu,\pi]} = -L^{\alpha}{}_{\rho[\pi}\, L^{\rho}{}_{\mu]\lambda} + \tfrac{1}{2}\, K^{\alpha}{}_{\lambda\pi\mu}. \tag{3.1}$$

As a system of partial differential equations for the $L^{\alpha}{}_{\beta\gamma}$, where the $K^{\alpha}{}_{\beta\gamma\lambda}$ are given, these equations are completely integrable† if and only if

$$L^{\alpha}{}_{\gamma[\mu,\pi,\omega]} = 0. \tag{3.2}$$

Substituting from the right-hand side of (3.1) into (3.2), we obtain, after algebraic simplification and use of the definition of the covariant derivative,

$$\tfrac{1}{2}\, K^{\alpha}{}_{\lambda[\nu\mu;\omega]} = 0. \tag{3.3}$$

If, now, we view $L^{\alpha}{}_{\beta\gamma}$ as actually given and $K^{\alpha}{}_{\beta\gamma\lambda}$ formed from it in accordance with (1.1), the integrability condition of (3.1) for $L^{\alpha}{}_{\beta\gamma}$ is identically satisfied, so that (3.3) is an identity, namely the celebrated Bianchi identity. Contracting (3.3) with respect to (α, λ) gives

$$K^{\alpha}{}_{\alpha[\nu\mu;\omega]} \equiv 0,$$

and hence, by (2.1),

$$A_{[\nu\mu;\omega]} \equiv 0, \tag{3.4}$$

which may equally well be written

$$A_{[\alpha\beta,\gamma]} \equiv 0 \tag{3.5}$$

because of the skew-symmetry of $A_{\alpha\beta}$. It should be noted that (3.5) has been *derived* from the definition of $A_{\alpha\beta}$ and the intrinsic properties of the curvature expression of the ground space $\mathfrak{A}$, a fact of which we shall later make significant use in the discussion of electromagnetic effects in the variant field theory. If the ground space under consideration happens to be a space with nonvanishing torsion tensor, the result stated in (3.4) and (3.5) ceases to hold. Instead,‡ we have

$$A_{[\alpha\beta;\gamma]} = X^{\rho}{}_{\cdot[\alpha\gamma}\, A_{\beta]\rho},$$

which will be shown to imply that in spaces of nonvanishing torsion there is no simple or direct road to equations representative of electromagnetic phenomena.

---

† J. A. Schouten: *op. cit.*, pp. 146, 153–155.
‡ *Ibid.*, p. 147.

Contracting (3.3) with respect to (α, μ) gives

$$0 = K^{\alpha}{}_{\lambda[\pi\mu;\alpha]} = \tfrac{1}{3}\,(K^{\alpha}{}_{\lambda\pi\mu;\alpha} + K^{\alpha}{}_{\lambda\mu\alpha;\pi} + K^{\alpha}{}_{\lambda\alpha\pi;\mu}),$$

and hence, by (2.2) and (2.3),

$$0 = K^{\alpha}{}_{\lambda\pi\mu;\alpha} + C_{\lambda\mu;\pi} + B_{\lambda\pi;\mu}.$$

By $(2.4)_1$, however, we have $B_{\lambda\pi} = -C_{\lambda\pi}$, and hence we obtain

$$\begin{aligned} 0 &= K^{\alpha}{}_{\lambda\pi\mu;\alpha} + C_{\lambda\mu;\pi} - C_{\lambda\pi;\mu} = K^{\alpha}{}_{\lambda\pi\mu;\alpha} + 2C_{\lambda[\mu;\pi]} \\ &= K^{\alpha}{}_{\lambda\pi\mu;\alpha} + 2R_{\lambda[\mu;\pi]} + A_{\lambda[\pi;\mu]} \end{aligned} \tag{3.6}$$

as the general case of the contraction of the Bianchi identity.

In a metric space, (1.7) and (3.6) give, after multiplication by $g^{\mu\lambda}$,

$$0 = J^{\alpha}{}_{\pi;\alpha} - J_{;\pi} + J^{\alpha}{}_{\pi;\alpha},$$

where $J \overset{\text{def}}{=} J_{\alpha\beta}\, g^{\alpha\beta}$. Hence in such spaces we have

$$G^{\mu}_{\cdot\,\pi;\mu} = 0, \tag{3.7}$$

where

$$G_{\alpha\beta} \overset{\text{def}}{=} J_{\alpha\beta} - \tfrac{1}{2}\, J\, g_{\alpha\beta} \tag{3.8}$$

are the components of the Einstein tensor (note that the identity $g^{\alpha\beta}{}_{;\gamma} = 0$ in an $\mathfrak{M}$ has been used).

Chapter III

The Axiomatic Theory of Variant Fields

WITHOUT FURTHER ADO, we proceed with axiomatic development of field space.

1. The Axioms of Field Space

Let $Q_{(\rho)}(\mathbf{x})$, $\rho = 1, \ldots, n$, be a collection of functions, with transformation properties under $'\mathbf{x} = \mathbf{f}(\mathbf{x})$ left unspecified. The axioms from which the variant field theory arises are contained in the following definition.

DEFINITION 1.1. *A general space is said to be a field space* $\mathfrak{F}$ *if and only if the following axioms are satisfied:*

F_1: $\mathfrak{F}$ *is an* $\mathcal{E}$.
F_2: *At each point of* $\mathfrak{F}$ *there is defined an affine connection* $L^{\alpha}{}_{\beta\gamma}$ ($\mathfrak{F}$ *is affinely connected*).
F_3: *The affine connection of* $\mathfrak{F}$ *is symmetric* ($\mathfrak{F}$ *is a space of identically vanishing torsion*).
F_4: *There is defined, at each point of* $\mathfrak{F}$, *a Lagrangian function*

$$\mathfrak{L}(A_{\alpha\beta}, R_{\alpha\beta}, L^{\alpha}{}_{\beta\gamma}, Q_{(\rho)}, Q_{(\rho),\mu}, \mathbf{x})$$

with no preassigned transformation properties under the transformations $'\mathbf{x} = \mathbf{f}(\mathbf{x})$ *and of class* C^2 *in its arguments. The field variables* $L^{\alpha}{}_{\beta\gamma}$ *and* $Q_{(\rho)}$ *are such that*

$$\delta \int_{D^*} \mathfrak{L}\, dV = 0 \tag{1.1}$$

for all variations $\delta L^{\alpha}{}_{\gamma\beta}$ *and* $\delta Q_{(\rho)}$ *vanishing on the boundary of* D^* *and satisfying* $\delta L^{\alpha}{}_{[\beta\gamma]} = 0$.

We shall say that any function $B(A_{\alpha\beta}, R_{\alpha\beta}, L^{\alpha}{}_{\beta\gamma}, Q_{(\rho)}, Q_{(\rho),\mu}, \mathbf{x})$ that is of class C^2 in its arguments is an *admissible* Lagrangian function.

Note. Although we could also include $L^{\alpha}{}_{\beta\gamma,\lambda}$ as an explicit argument of $\mathfrak{L}$, in addition to its implicit occurrence through $A_{\alpha\beta}$ and $R_{\alpha\beta}$, this added complexity is found to be unnecessary.

Some comment is in order here on the degree of constraint represented by the four axioms of field space. The condition F_1 is an extremely mild constraint, and ensures only that the normal processes of analysis are well defined locally. Axiom F_2 is the minimal constraint for which comparison of local vector fields with respect to parallelism is well defined and for which the normal notions of geometry are manifest. Axiom F_3, which assures the commutativity of invariant coördinate-segment addition, might seem on first examination to be a somewhat stronger constraint than the former two conditions. It has been shown by Schouten,† however, that—as far as curvature considerations are concerned—any result that can be obtained in a space with torsion can equally well be obtained in a subsidiary torsion-free space with a skew 2-form. Since our considerations will center on curvature expressions, we see that F_3 is also a mild constraint. Axiom F_4 is a very strong constraint. It requires not only the existence of a Lagrangian function, but also that equation (1.1) be satisfied with the components of affine connection as independent variables in the variation process; F_4 thus leads to a system of equations for the determination of the components of affine connection and thereby assigns a particular structure to $\mathfrak{F}$. Axiom F_4 is the basic generating axiom in the theory of variant fields.

There is no direct basis for F_4 other than the fact that almost all previous theories of fields and the theories of classical and quantum physics can be obtained from suitably chosen variational statements. In the previous theories, however, the Lagrangian function has usually been assumed to have particular transformation properties and to depend on its various arguments in very specific ways, or to be assigned uniquely. We make no such assumptions in the considerations to follow. *One of the main features of the variant field theory is that it holds for arbitrary Lagrangian functions and spans the set of all theories resulting from particularizations of that function.*

† J. A. Schouten: *op. cit.*, p. 184.

2. Variational Considerations and the Variant Field Equations

Now that we have set down the axioms of field space, the next logical step is to investigate their implications. Let us define the quantities $I^{\alpha\beta}$ and $H^{\alpha\beta}$ by

$$I^{\alpha\beta} = \mathfrak{L}_{,R\alpha\beta}, \tag{2.1}$$

$$H^{\alpha\beta} = \mathfrak{L}_{,A\alpha\beta}; \tag{2.2}$$

these are symmetric and skew-symmetric, respectively, as a result of the respective symmetry and skew-symmetry of $R_{\alpha\beta}$ and $A_{\alpha\beta}$. Using equations (II-2.12) and (II-2.13), and assuming that $R_{\alpha\beta}$, $A_{\alpha\beta}$, and $Q_{(\rho),\mu}$ are elements of class C^1 so that $I^{\alpha\beta}$, $H^{\alpha\beta}$, and $\mathfrak{L}_{,Q(\rho),\mu}$ are of class C^1 by the assumed continuity of $\mathfrak{L}$, we find through direct computation that

$$\begin{aligned}\delta \int_{D^*} \mathfrak{L}\, dV &= \int_{D^*} [\mathfrak{L}_{,Q_{(\rho)}} - (\mathfrak{L}_{,Q_{(\rho),\xi}})_{,\xi}]\, \delta Q_{(\rho)}\, dV \\ &+ \int_{D^*\ominus D} [H^{\alpha\beta}\,(\delta L^{\sigma}{}_{\sigma\beta}\, N_\alpha - \delta L^{\sigma}{}_{\sigma\alpha}\, N_\beta) + (\mathfrak{L}_{,Q_{(\rho),\xi}})\, \delta Q_{(\rho)}\, N_\xi \\ &\qquad + I^{\alpha\beta}\,(\tfrac{1}{2}\, \delta L^{\epsilon}{}_{\epsilon\beta}\, N_\alpha + \tfrac{1}{2}\, \delta L^{\epsilon}{}_{\epsilon\alpha}\, N_\beta - \delta L^{\epsilon}{}_{\alpha\beta}\, N_\epsilon)]\, dS \\ &+ \int_{D^*} [-I^{\beta\epsilon}{}_{,\epsilon}\, \delta^\gamma_\alpha + I^{\beta\gamma}{}_{,\alpha} - I^{\beta\gamma}\, L^{\epsilon}{}_{\epsilon\alpha} + I^{\epsilon\gamma}\, L^{\beta}{}_{\epsilon\alpha} + I^{\epsilon\beta}\, L^{\gamma}{}_{\epsilon\alpha} \\ &\qquad - I^{\epsilon\rho}\, L^{\beta}{}_{\epsilon\rho}\, \delta^\gamma_\alpha + 2H^{\gamma\epsilon}{}_{,\epsilon}\, \delta^\beta_\alpha + \partial_{L^\alpha{}_{\beta\gamma}} \mathfrak{L}]\, \delta L^{\alpha}{}_{\beta\gamma}\, dV,\end{aligned}$$

where $\partial_{L^\alpha{}_{\beta\gamma}}\mathfrak{L}$ stands for $\partial\mathfrak{L}/\partial L^{\alpha}{}_{\beta\gamma}$ evaluated for constant $A_{\alpha\beta}$ and $R_{\alpha\beta}$. This result can be simplified by interpreting $I^{\alpha\beta}$ *as if* it were a contravariant tensor of rank two and weight one and then applying the operation of covariant differentiation with respect to $L^{\alpha}{}_{\beta\gamma}$. The quantity $I^{\alpha\beta}$ is not necessarily a tensor density, however, since the Lagrangian function has been left general with respect to its transformation properties. Hence we may not represent such an operation by the normal semicolon (covariant) derivative. To avoid unnecessary confusion, we define a new derivative, the colon derivative,

$$I^{\alpha\beta}{}_{:\gamma} \overset{\text{def}}{=} I^{\alpha\beta}{}_{,\gamma} + L^{\alpha}{}_{\eta\gamma}\, I^{\eta\beta} + L^{\beta}{}_{\eta\gamma}\, I^{\eta\alpha} - L^{\eta}{}_{\eta\gamma}\, I^{\alpha\beta}, \tag{2.3}$$

which is just the covariant derivative *if* $I^{\alpha\beta}$ should turn out to be a tensor density. With this operational agreement, we have

$$\delta \int_{D^*} \mathfrak{L}\, dV = \int_{D^*} [\mathfrak{L}_{,Q_{(\rho)}} - (\mathfrak{L}_{,Q_{(\rho),\xi}})_{,\xi}]\, \delta Q_{(\rho)}\, dV$$

$$+ \int_{D^*} [\partial_{L^\alpha{}_{\beta\gamma}} \mathfrak{L} + I^{\beta\gamma}{}_{:\alpha} - \delta^{(\gamma}_{\alpha} I^{\beta)\epsilon}{}_{:\epsilon} + 2\delta^{(\beta}_{\alpha} H^{\gamma)\epsilon}{}_{,\epsilon}]\, \delta L^\alpha{}_{\beta\gamma}\, dV$$

$$+ \int_{D^* \ominus D} [2H^{\alpha\beta}\, \delta L^\epsilon{}_{\epsilon[\beta} N_{\alpha]} + I^{\alpha\beta}\, (\delta L^\epsilon{}_{\epsilon(\alpha} N_{\beta)} - \delta L^\epsilon{}_{\alpha\beta} N_\epsilon)$$

$$+ \mathfrak{L}_{,Q_{(\rho),\xi}}\, \delta Q_{(\rho)}\, N_\xi]\, dS, \tag{2.4}$$

on making use of the conditions $L^\alpha{}_{[\beta\gamma]} = 0$ and $\delta L^\alpha{}_{[\beta\gamma]} = 0$ of axioms F_3 and F_4. Thus, annulling $\delta L^\alpha{}_{\beta\gamma}$ and $\delta Q_{(\rho)}$ on the boundary of D^* in conformity with axiom F_4, we obtain the following conditions for satisfaction of equation (1.1) by application of well-known results from the calculus of variations:

$$I^{\beta\gamma}{}_{:\alpha} - \delta^{(\beta}_{\alpha} I^{\gamma)\eta}{}_{:\eta} + \partial_{L^\alpha{}_{\beta\gamma}} \mathfrak{L} + 2\delta^{(\beta}_{\alpha} H^{\gamma)\eta}{}_{,\eta} = 0, \tag{2.5}$$

$$\mathfrak{L}_{,Q_{(\rho)}} - (\mathfrak{L}_{,Q_{(\rho),\xi}})_{,\xi} = 0, \quad \rho = 1, \ldots, n. \tag{2.6}$$

Equations (2.5) and (2.6) are $40 + n$ equations for the determination of the $40 + n$ field variables $L^\alpha{}_{\beta\gamma}$ and $Q_{(\rho)}$; they are the *variant field equations*. (Note that $H^{\alpha\beta}$ and $I^{\alpha\beta}$ are known functions of $L^\alpha{}_{\beta\gamma}$, $Q_{(\rho)}$, and their derivatives, once the Lagrangian function $\mathfrak{L}$ has been given (see (2.1) and (2.2)). In this sense, $\mathfrak{L}$ acts as a constitutive function for the $I^{\alpha\beta}$ and $H^{\alpha\beta}$ fields.) We have thus proved the following basic result.

THEOREM 2.1. *If $R_{\alpha\beta}$, $A_{\alpha\beta}$, and $Q_{(\rho),\mu}$ are of class C^1, Axiom F_4 is satisfied if and only if $L^\alpha{}_{\beta\gamma}$ and $Q_{(\rho)}$ are such that the variant field equations* (2.5) *and* (2.6) *are satisfied in every D^* contained in $\mathfrak{F}$.*

This result now allows us to characterize field space in the following manner.

THEOREM 2.2. *Let $\mathfrak{S}$ be an affinely connected $\mathcal{E}$ with vanishing torsion tensor, and let $R_{\alpha\beta}$, $A_{\alpha\beta}$, and $Q_{(\rho),\mu}$ be elements of class C^1. Then $\mathfrak{S}$ is an $\mathfrak{F}$ if and only if*

$L^\alpha{}_{\beta\gamma}$ and $Q_{(\rho)}$ are such that the variant field equations (2.5) *and* (2.6) *are satisfied for every D^* contained in $\mathcal{S}$.*

A particular decomposition of the variant field equation (2.5), which we shall often find convenient, is given in the following theorem.

THEOREM 2.3. *The equations* (2.5) *are equivalent to the system*

$$H^{\alpha\eta}{}_{,\eta} = \tfrac{1}{10}(3I^{\alpha\eta}{}_{:\eta} - 2\partial_{L^\eta{}_{\eta\alpha}}\mathfrak{L}), \tag{2.7}$$

$$I^{\beta\gamma}{}_{:\alpha} - \tfrac{2}{5}\delta^{(\beta}_{\alpha} I^{\gamma)\eta}{}_{:\eta} = P^{\beta\gamma}_{\cdot\cdot\alpha}, \tag{2.8}$$

where

$$P^{\beta\gamma}_{\cdot\cdot\alpha} \overset{\text{def}}{=} (\tfrac{2}{5}\delta^{(\beta}_{\alpha}\partial_{L^\eta{}_{\gamma)\eta}} - \partial_{L^\alpha{}_{\beta\gamma}})\,\mathfrak{L}. \tag{2.9}$$

Proof. Contracting (2.5) with respect to either (α, β) or (α, γ) gives (2.7). The elimination of $H^{\alpha\eta}{}_{,\eta}$ between (2.5) and (2.7) then leads to (2.8) under the definition (2.9).

Note. If we had not assumed that the torsion tensor vanishes, the equation that would replace (2.5), when contracted with respect to (α, β) and (α, γ), would give two different right-hand sides to (2.7). The equation (2.5) would then be replaced by the system composed of equations similar to (2.7)–(2.9) together with the equation that would result from equating the two different right-hand sides of the equation replacing (2.7). There would thus be a total of $68 + n$ equations for the determination of only $64 + n$ variables.

It should be noted that the variant field equations make statements concerning certain operational functions of the field variables $L^\alpha{}_{\beta\gamma}$ and $Q_{(\rho)}$. In order to see this, we rewrite the field equations and their subsidiary relations in the following form:

$$\left.\begin{aligned} H^{\alpha\beta} &= \mathfrak{L}_{,A_{\alpha\beta}}, \\ I^{\alpha\beta} &= \mathfrak{L}_{,R_{\alpha\beta}}, \end{aligned}\right\} \tag{2.10}$$

$$\left.\begin{aligned} Z^{(\rho)} &= \mathfrak{L}_{,Q_{(\rho)}}, \qquad (\rho = 1, \ldots, n) \\ W^{(\rho)\eta} &= \mathfrak{L}_{,Q_{(\rho),\eta}}, \end{aligned}\right\} \tag{2.11}$$

$$P^{\beta\gamma}_{\cdot\cdot\alpha} = (\tfrac{2}{5}\delta^{(\beta}_{\alpha}\partial_{L^\eta{}_{\gamma)\eta}} - \partial_{L^\alpha{}_{\beta\gamma}})\,\mathfrak{L}, \tag{2.12}$$

$$H^{\alpha\eta}{}_{,\eta} = \tfrac{1}{10}\left(3I^{\alpha\eta}{}_{:\eta} - 2\partial_{L^{\eta}{}_{\eta\alpha}}\mathfrak{L}\right),$$

$$I^{\beta\gamma}{}_{;\alpha} = \tfrac{2}{5}\,\delta^{(\beta}_{\alpha}\, I^{\gamma)\eta}{}_{:\eta} + P^{\beta\gamma}_{\cdot\cdot\alpha}\,, \tag{2.13}$$

$$Z^{(\rho)} = W^{(\rho)\eta}{}_{,\eta}.$$

The field equations (2.13) are thus seen to relate the quantities

$$H^{\alpha\beta},\quad I^{\alpha\beta},\quad P^{\beta\gamma}_{\cdot\cdot\alpha},\quad Z^{(\rho)},\quad W^{(\rho)\eta},\qquad \text{and}\qquad \partial_{L^{\eta}{}_{\eta\alpha}}\mathfrak{L},$$

which are subsidiary fields defined over field space by (2.10)–(2.12) in terms of the Lagrangian function. These fields are functions of $A_{\alpha\beta}$, $R_{\alpha\beta}$, $L^{\alpha}{}_{\beta\gamma}$, $Q_{(\rho)}$, $Q_{(\rho),\eta}$, and $\mathbf{x}$, by virtue of the assumed arguments of the Lagrangian function, but are not necessarily tensorial fields since the Lagrangian function has not been assumed to have any preassigned transformation properties. Hence the variant field equations make statements concerning certain collections of subsidiary fields. These statements are not necessarily tensorial, and the collection of subsidiary fields involved in these statements are general functional derivatives of the Lagrangian function.

3. Boundary Conditions and Discontinuity Surfaces

The derivation of the field equations given in Section 2 assumed that all the field variables were continuous with continuous derivatives in the D^* under consideration. We shall now relax this condition.

Let the domain D be divided into two parts D^+ and D^- by a 3-surface π, so that $D^* = (D^+ \cup D^-)^*$. We assume that all field variables are continuously differentiable in D^+ and D^- and have limiting values on π. If we represent a typical field variable by Δ, then Δ^+ and Δ^- will be used to represent the limiting values of Δ as π is approached from D^+ and D^-, respectively. The discontinuity in Δ across π is thus given by

$$[\![\Delta]\!] \stackrel{\text{def}}{=} \Delta^+ - \Delta^-. \tag{3.1}$$

If Δ is a vector-ordered quantity, say Δ^α, which is continuously differentiable in D^+ and D^-, then the divergence theorem yields

$$\int_{D^*} \Delta^{\alpha}{}_{,\alpha}\, dV + \int_{D^*\cap\pi} [\![\Delta^\alpha]\!]\, J_\alpha\, d\pi = \int_{D^*\ominus(D^+\cup D^-\cup\pi)} \Delta^\alpha N_\alpha\, dS, \tag{3.2}$$

where $J_\alpha\, d\pi$ is the symbolic vector-ordered element of surface corresponding to an element of π intersecting the closure of D^-, and hence is oriented from D^- to D^+.

Let $\mathfrak{F}$ be a field space. Then in a manner directly similar to that used in deriving (2.4) we use (3.2) to obtain

$$\begin{aligned}\delta \int_{D^*} \mathfrak{L}\, dV = &\int_{D^*} (Z^{(\rho)} - W^{(\rho)\xi}{}_{,\xi})\, \delta Q_{(\rho)}\, dV \\ &+ \int_{D^*} (\partial_{L^\alpha{}_{\beta\gamma}} \mathfrak{L} + I^{\beta\gamma}{}_{:\alpha} - \delta^\gamma_\alpha I^{\beta\epsilon}{}_{:\epsilon} + 2\delta^\beta_\alpha H^{\gamma\epsilon}{}_{,\epsilon})\, \delta L^\alpha{}_{\beta\gamma}\, dV \\ &+ \int_{D^* \ominus (D^+ \cup D^- \cup \pi)} [2H^{\alpha\beta}\, \delta L^\epsilon{}_{\epsilon[\beta} N_{\alpha]} + I^{\alpha\beta} (\delta L^\epsilon{}_{\epsilon(\alpha} N_{\beta)} \\ &\qquad - \delta L^\epsilon{}_{\alpha\beta} N_\epsilon) + W^{(\rho)\xi}\, \delta Q_{(\rho)} N_\xi]\, dS \\ &- \int_{D^* \cap \pi} [\![2H^{\alpha\beta}\, \delta L^\epsilon{}_{\epsilon[\beta} J_{\alpha]} + I^{\alpha\beta}(\delta L^\epsilon{}_{\epsilon(\alpha} J_{\beta)} - \delta L^\epsilon{}_{\alpha\beta} J_\epsilon) \\ &\qquad + W^{(\rho)\xi}\, \delta Q_{(\rho)} J_\xi]\!]\, d\pi. \end{aligned} \tag{3.3}$$

If we set

$$Y^{\beta\gamma\mu}_\alpha \stackrel{\text{def}}{=} \delta^\beta_\alpha (2H^{\mu\gamma} + I^{\mu\gamma}) - I^{\beta\gamma}\, \delta^\mu_\alpha, \tag{3.4}$$

(3.3) becomes

$$\begin{aligned}\delta \int_{D^*} \mathfrak{L}\, dV = &\int_{D^*} (Z^{(\rho)} - W^{(\rho)\xi}{}_{,\xi})\, \delta Q_{(\rho)}\, dV \\ &+ \int_{D^*} (\partial_{L^\alpha{}_{\beta\gamma}} \mathfrak{L} + I^{\beta\gamma}{}_{:\alpha} - \delta^\gamma_\alpha I^{\beta\epsilon}{}_{:\epsilon} + 2\delta^\beta_\alpha H^{\gamma\epsilon}{}_{,\epsilon})\, \delta L^\alpha{}_{\beta\gamma}\, dV \\ &+ \int_{D^* \ominus (D^+ \cup D^- \cup \pi)} (Y^{\beta\gamma\mu}_\alpha\, \delta L^\alpha{}_{\beta\gamma} N_\mu + W^{(\rho)\xi}\, \delta Q_{(\rho)} N_\xi)\, dS \\ &- \int_{D^* \cap \pi} [\![Y^{\beta\gamma\mu}_\alpha\, \delta L^\alpha{}_{\beta\gamma} + W^{(\rho)\mu}\, \delta Q_{(\rho)}]\!]\, J_\mu\, d\pi. \end{aligned} \tag{3.5}$$

By axiom F_4, $\delta L^\alpha{}_{\beta\gamma}$ is symmetric, and hence

$$\delta \int_{D^*} \mathfrak{L}\, dV = \int_{D^*} (Z^{(\rho)} - W^{(\rho)\xi}{}_{,\xi})\, \delta Q_{(\rho)}\, dV$$

$$+ \int_{D^*} (\partial_{L^\alpha{}_{\beta\gamma}} \mathfrak{L} - I^{\beta\gamma}{}_{:\alpha} - \delta^{(\gamma}_{\alpha} I^{\beta)\epsilon}{}_{:\epsilon} + 2\delta^{(\beta}_{\alpha} H^{\gamma)\epsilon}{}_{,\epsilon})\, \delta L^\alpha{}_{\beta\gamma}\, dV$$

$$+ \int_{D^* \ominus (D^+ \cup D^- \cup \pi)} (Y^{(\beta\gamma)\mu}_{\alpha}\, \delta L^\alpha{}_{\beta\gamma} + W^{(\rho)\mu}\, \delta Q_{(\rho)})\, N_\mu\, dS$$

$$+ \int_{D^* \cap \pi} [\![Y^{(\beta\gamma)\mu}_{\alpha}\, \delta L^\alpha{}_{\beta\gamma} + W^{(\rho)\mu}\, \delta Q_{(\rho)}]\!]\, J_\mu\, d\pi. \tag{3.6}$$

Applying the remaining conditions of axiom F_4 with regard to the vanishing of the variations on the boundary, and requiring $\delta L^\alpha{}_{\beta\gamma}$ and $\delta Q_{(\rho)}$ to be continuous on π, we obtain the following result.

THEOREM 3.1. *Let the domain D in field space be divided into two parts D^+ and D^- by a 3-surface π, and suppose that some of the field variables have discontinuous derivatives*† *on π, but that $\delta L^\alpha{}_{\beta\gamma}$ are continuous. Then the field variables must satisfy the field equations* (2.13) *at all points in $D^* \ominus \pi$; and the equations*

$$[\![Y_\alpha{}^{(\beta\gamma)\mu}]\!]\, J_\mu = 0, \tag{3.7}$$

$$[\![W^{(\rho)\mu}]\!]\, J_\mu = 0, \tag{3.8}$$

must be satisfied at each point of $D^* \cap \pi$, where $Y^{\beta\gamma\mu}_{\alpha}$ is defined by (3.4).

Expanding $Y^{(\beta\gamma)\mu}_{\alpha}$ by use of (3.4), we have

$$Y^{(\beta\gamma)\mu}_{\alpha} = \delta^{(\beta}_{\alpha} I^{\gamma)\mu} - 2\delta^{(\beta}_{\alpha} H^{\gamma)\mu} - I^{\beta\gamma}\, \delta^{\mu}_{\alpha}. \tag{3.9}$$

Thus we obtain

$$[\![Y^{(\alpha\gamma)\mu}_{\alpha}]\!]\, J_\mu = \tfrac{1}{2} [\![3I^{\gamma\mu} - 10H^{\gamma\mu}]\!]\, J_\mu = 0, \tag{3.10}$$

so that

$$[\![H^{\gamma\mu}]\!]\, J_\mu = +\tfrac{3}{10} [\![I^{\gamma\mu}]\!]\, J_\mu. \tag{3.11}$$

Substituting (3.11) into (3.7) we have the following result, similar to Theorem 2.2.

† If $[\![L^\alpha{}_{\beta\gamma,\delta}]\!] \neq 0$, then $[\![I^{\alpha\beta}]\!] \neq 0$ or $[\![H^{\alpha\beta}]\!] \neq 0$ for some (α, β).

THEOREM 3.2. *The boundary conditions* (3.7) *are equivalent to the system*

$$\begin{aligned} [\![H^{\gamma\mu}]\!]\, J_\mu &= \tfrac{3}{10}\, [\![I^{\gamma\mu}]\!]\, J_\mu, \\ [\![I^{\beta\gamma}\, \delta^\mu_\alpha - \tfrac{2}{5}\, \delta^{(\beta}_\alpha\, I^{\gamma)\mu}]\!]\, J_\mu &= 0. \end{aligned} \tag{3.12}$$

4. Equations of Maxwellian Form as Natural Consequences of the Axioms of Field Space

A fundamental property of field space will be proved in this section, namely, that equations of *Maxwellian form* are a natural and direct consequence of the axioms that characterize field space.

We first recall the form of Maxwell's equations for a linear medium as they appear in 4-dimensional notation:†

$$f_{(\alpha\beta)} = 0, \qquad f_{[\alpha\beta,\gamma]} = 0,$$

$$F^{(\alpha\beta)} = 0, \qquad F^{\alpha\eta}{}_{,\eta} = S^\alpha,$$

$$F^{\alpha\beta} = G^{\alpha\beta\gamma\lambda}\, f_{\gamma\lambda},$$

where S^α is the 4-current, $f_{\alpha\beta}$ the electromagnetic field tensor, $F^{\alpha\beta}$ the current potentials, and $G^{\alpha\beta\gamma\lambda}$ a quantity representing the properties of the medium. With these equations as a guide, we lay down the following definition.

DEFINITION 4.1. *A system of equations is said to be of Maxwellian form if and only if it can be written as*

$$\begin{aligned} &f_{(\alpha\beta)} = 0, \quad f_{[\alpha\beta,\gamma]} = 0, \quad f_{\alpha\beta} \text{ a tensor}, \\ &F^{(\alpha\beta)} = 0, \quad F^{\alpha\eta}{}_{,\eta} = S^\alpha, \quad F^{\alpha\beta} = U(f_{\eta\mu}, \ldots)_{,f_{\alpha\beta}}. \end{aligned} \tag{4.1}$$

Under these conditions $f_{\alpha\beta}$ is referred to as the Maxwellian field tensor, S^α is referred to as the Maxwellian current, $F^{\alpha\beta}$ are referred to as the Maxwellian current potentials, and $U(f_{\eta\mu}, \ldots)$ is referred to as the constitutive function for the Maxwellian field.‡

† See J. L. Synge: *Relativity: The General Theory*, North-Holland Publishing Co., Amsterdam (1960), p. 356.

‡ Equations of Maxwellian form are not necessarily Maxwell's equations for the electromagnetic field, although they are identical in formal structure. Such systems of equations have been considered by many authors for the purpose of providing a description of possible generalizations of the electromagnetic field. The concept of the

Let $\mathfrak{F}$ be a field space. The existence of an affine connection is ensured by axiom F_2, and hence the curvature tensor exists and is well defined. The contraction $A_{\alpha\beta}$ of the curvature tensor thus exists and is skew-symmetric (that is, we have $A_{(\alpha\beta)} = 0$). By F_3, the torsion tensor $X^{\alpha}{}_{\beta\gamma}$ vanishes, and hence we have $A_{[\alpha\beta;\gamma]} = 0$ as a consequence of the differential identities satisfied by the curvature tensor (see II-3). Hence the equation

$$A_{[\alpha\beta;\gamma]} = 0$$

always holds in $\mathfrak{F}$. Since $A_{\alpha\beta}$ is skew-symmetric, it is evident upon expanding

$$A_{[\alpha\beta;\gamma]} = 0$$

that this equation is equivalent to

$$A_{[\alpha\beta,\gamma]} = 0.$$

We thus have the following results under axioms F_2 and F_3:

$$A_{(\alpha\beta)} = 0, \qquad A_{[\alpha\beta,\gamma]} = 0, \qquad A_{\alpha\beta} \text{ is a tensor.}$$

Axiom F_4 ensures the existence of a Lagrangian function, and hence

$$H^{\alpha\beta} = \mathfrak{L}_{,A_{\alpha\beta}}$$

exists in $\mathfrak{F}$ and is skew-symmetric over (α, β). We have also seen, by Theorem 2.1 and 2.2, that axiom F_4 implies

$$H^{\alpha\eta}{}_{,\eta} = S^{\alpha},$$

where S^{α} denotes the right-hand side of (2.7), that is,

$$S^{\alpha} = \tfrac{1}{10}\,(3I^{\alpha\eta}{}_{:\eta} - 2\partial_{L^{\eta}{}_{\eta\alpha}}\mathfrak{L}).$$

Combining these results and using Definition 3.1, we have the following result.

---

field $f_{\alpha\beta}$, rather than the charge and current distributions, as the seat of electromagnetic phenomena can be credited to G. Mie: "Grundlagen einer Theorie der Materie," *Ann. Physik*, Series 4, **37** (1912), pp. 511–534; **39** (1912), pp. 1–40; **40** (1913), pp. 1–66. An excellent exposition of Mie's considerations is given by H. Weyl: *Space–Time–Matter*, Dover Publications, Inc., New York (1950), pp. 206–217, and will be seen to have many properties in common with our results, particularly with respect to the constitutive equation obtained for the Maxwellian current S^{α}.

THEOREM 4.1. *The following system of equations of Maxwellian form is a direct consequence of the axioms of field space:*

$$A_{(\alpha\beta)} = 0, \qquad A_{[\alpha\beta,\gamma]} = 0, \tag{4.2}$$

$$H^{(\alpha\beta)} = 0, \qquad H^{\alpha\eta}{}_{,\eta} = S^{\alpha}, \qquad H^{\alpha\beta} = \mathfrak{L}_{,A_{\alpha\beta}}. \tag{4.3}$$

Thus $A_{\alpha\beta}$ is the Maxwellian field tensor, $H^{\alpha\beta}$ are the Maxwellian current potentials, and the Lagrangian function is the constitutive function for this Maxwellian field. In addition, S^{α} is the Maxwellian current, with constitutive equation

$$S^{\alpha} = \tfrac{1}{10}\,(3I^{\alpha\eta}{}_{:\eta} - 2\partial_{L^{\eta}{}_{\eta\alpha}}\mathfrak{L}). \tag{4.4}$$

It may seem strange that we have been able to obtain equations of Maxwellian form using only the axioms of field space, *without* specializing the Lagrangian function, and without assuming any metric structure for the space. We must note that equations of Maxwellian form are of a topological nature, expressing as they do the conservation of the current S^{α} and the current potentials, and the existence and solenoidal character of the vector potential.† In fact, it can be shown that the general system of Maxwell's equations for the electromagnetic field can be cast in such a form that they are completely independent of metric geometric considerations.‡

Since $A_{\alpha\beta}$ is a tensor, (4.2) ensures the existence of a vector f_{η} such that

$$f_{[\alpha,\beta]} = A_{\alpha\beta}. \tag{4.5}$$

The vector f_{η}, so defined, will be referred to as the *Maxwellian vector potential.*

Expanding (4.5) by means of (II-2.5), we have

$$f_{[\alpha,\beta]} = 2L^{\eta}{}_{\eta[\beta,\alpha]} = -2L^{\eta}{}_{\eta[\alpha,\beta]}.$$

Although the quantities $L^{\eta}{}_{\eta\alpha}$ are not the components of a vector under admissible coördinate transformations, they do transform as components of a vector under the subgroup of linear coördinate transformations. Hence, under the linear group we may take $-1/2\, L^{\eta}{}_{\eta\alpha}$ as the Maxwellian vector potential. Under general coördinate transformations, we shall

† For a particularly elegant presentation of electromagnetic theory from the standpoint of its topological statements, see C. Truesdell and R. Toupin: "The Classical Field Theories," *Handbuch der Physik*, Band III/1, Springer-Verlag, Berlin (1960), pp. 660–700.

‡ D. van Dantzig: "The Fundamental Equations of Electromagnetism, Independent of Metric Geometry," *Cambridge Phil. Soc. Proc.*, **30** (1935), pp. 421–427.

consider $L^{\eta}{}_{\eta\alpha}$ as the *components of potential* of the Maxwellian field tensor $A_{\alpha\beta}$. In many respects, treating $L^{\eta}{}_{\eta\alpha}$ as a potential form for $A_{\alpha\beta}$ is more fundamental than considering a vector potential. This is due to the fact that we can find coördinate systems in which $L^{\eta}{}_{\eta\alpha}$ will vanish at a preassigned point (for example, at the origin of a normal coördinate system†) but will not vanish for other choices of coördinate systems. The freedom of choosing $L^{\eta}{}_{\eta\alpha}$ to vanish at a preassigned point is equivalent to the freedom of choice of a reference point and a reference value for the potential.

Equations $(4.3)_2$ yield, by a direct computation, the topological conservation law for the Maxwellian current, namely

$$S^{\alpha}{}_{,\alpha} = 0. \tag{4.6}$$

If $\mathfrak{L}$ is a scalar density, then $H^{\alpha\beta}$ is a contravariant tensor density, and the above topological conservation law becomes the covariant conservation law

$$S^{\alpha}{}_{;\alpha} = 0.$$

In this section, results having the *formal structure* of Maxwell's electrodynamics have been obtained as a direct consequence of the axioms of field space, by a simple separation of the tensor $C_{\alpha\beta}$ into the tensors $A_{\alpha\beta}$ and $R_{\alpha\beta}$, without any assumptions about the form of the constitutive equations describing the physical fields. (It is this separation that yielded the Maxwellian field tensor $A_{\alpha\beta}$ and the basic equations (2.7).) This is as it should be. If we are to obtain representations of electromagnetic phenomena from a field theory, the basic equations representative of electromagnetic phenomena (namely, equations of Maxwellian form) should be independent of the particular field we wish to investigate; that is to say, they should be independent of the particular constitutive functions used. We must, however, pick the appropriate constitutive functions if theoretical predictions are to match experimental results.

Since spaces with intrinsic torsion, that is, spaces for which $X^{\alpha}{}_{\beta\gamma} \neq 0$, are basic in the unified field theories of Einstein, Hlavatý, and Schrödinger,‡ it is of interest to examine what results when we relax axiom F_3. First, as noted in the previous section, we do not directly obtain an equation of the form

$$H^{\alpha\eta}{}_{,\eta} = S^{\alpha}$$

† T. Y. Thomas: *op. cit.*, Chap. V.

‡ E. Schrödinger: "The Final Affine Field Laws," *Proc. Roy. Irish Ac.*, **A51**, pp. 163–171, 205–216; **A52**, pp. 1–9 (1947/1948).

by contracting the field equations. The equations that would replace the preceding one result in two different expressions for S^α, since the equations equivalent to (2.5) are then not symmetric in the two free superscripts. Hence these two expressions must be equated, yielding four additional equations. We are thus unable to obtain equations of Maxwellian form in any simple and direct manner. In addition, if the torsion tensor does not vanish, the differential identities of the curvature tensor yield

$$A_{[\alpha\beta;\gamma]} = 2X^{\rho}{}_{[\gamma\alpha} A_{\beta]\rho}$$

instead of

$$A_{[\alpha\beta;\gamma]} = 0.$$

Because of the skew-symmetry of the tensor $A_{\alpha\beta}$, the above equation is still equivalent to

$$A_{[\alpha\beta,\gamma]} = 0,$$

but many of the formal results of Maxwell's electrodynamics no longer hold. For example, we have seen that there exists a vector f_η such that

$$A_{\alpha\beta} = f_{[\alpha,\beta]},$$

as a consequence of

$$A_{[\alpha\beta,\gamma]} = 0.$$

Expanding $f_{[\alpha;\beta]}$ by the definition of the covariant derivative, we have

$$f_{[\alpha;\beta]} = f_{[\alpha,\beta]} - A^{\sigma}{}_{[\beta\alpha]} f_\sigma = f_{[\alpha,\beta]} - X^{\sigma}{}_{\beta\alpha} f_\sigma,$$

in accordance with the definition of the torsion tensor ($X^{\sigma}{}_{\beta\alpha} = A^{\sigma}{}_{[\beta\alpha]}$), and hence

$$f_{[\alpha,\beta]} = f_{[\alpha;\beta]} + X^{\sigma}{}_{\beta\alpha} f_\sigma.$$

Substituting this result into

$$A_{\alpha\beta} = f_{[\alpha,\beta]},$$

we then have

$$A_{\alpha\beta} = f_{[\alpha;\beta]} + X^{\sigma}{}_{\beta\alpha} f_\sigma.$$

Hence, if the torsion tensor does not vanish, the potential representation of $A_{\alpha\beta}$ in covariant form depends not only on the covariant derivative of f_α but also on f_α and the torsion tensor. More important, we see that even if

$$f_{[\alpha;\beta]} = 0,$$

the field $A_{\alpha\beta}$ is nonzero if

$$X^\sigma{}_{\alpha\beta} f_\sigma \neq 0.$$

On the other hand, if the torsion tensor vanishes, then the potential representation of $A_{\alpha\beta}$ depends only on the covariant derivative of f_α, as is customary.

It is of interest to examine what happens in a space that is assumed to be metric from the start.

THEOREM 4.2. *If a space $\mathcal{G}$ is assumed a priori to be a metric space, $\mathfrak{M}$, it is impossible to obtain equations of Maxwellian form from a variational principle with Lagrangian function depending* only *on contractions of the curvature tensor, the metric tensor $g_{\alpha\beta}$, and the affine connection; rather, electromagnetic phenomena must be introduced by the inclusion of additional field variables.*

Proof. By Theorem II-2.2, we have

$$A_{\alpha\beta} = 0$$

if the space is metric. The hypothesized arguments of the Lagrangian function then are all symmetric, since

$$C_{[\alpha\beta]} = -B_{[\alpha\beta]} = \tfrac{1}{2} A_{\beta\alpha} = 0.$$

Thus, among the assumed arguments of the Lagrangian function, there is no skew-symmetric tensor available to replace the tensor $A_{\alpha\beta}$, and hence additional skew-symmetric field variables must be introduced in order for a variational principle to lead to field equations of Maxwellian form; this establishes the theorem.

Theorem 4.2 lies at the heart of the problem of constructing a field theory that, from a common geometric basis, naturally admits electromagnetic and gravitational phenomena.

The variational theory of Lanczos† might appear to contradict the

† C. Lanczos: "Electricity and General Relativity," *Rev. Mod. Phys.*, **29** (1957), pp. 337–350.

foregoing results on first reading. This is not the case, however, for the vector ϕ_μ introduced by Lanczos through a "canonical transformation" adds to the Lagrangian function of his theory just the skew-symmetric tensor required to yield equations of Maxwellian form. The "already unified" theory of Rainich–Misner–Wheeler† might also seem to contradict the above theorem. But again this is not so, for the assumptions that must be made in order to obtain the "already unified" theory are equivalent to setting the current vector equal to zero at all ordinary points of the field and equating the Einstein tensor to an a priori given momentum-energy tensor of the electromagnetic field in vacuum. One thus loses equations (4.3) and can recover their content only by assuming that the "already unified" electromagnetic field arises from singularities in the solutions of the field equations.‡ In effect, this is a case of robbing Peter to pay Paul, and all that really results is a very particular kind of electromagnetism veiled as geometry.

The important identification of the curvature tensor $A_{\alpha\beta}$ with the Maxwellian field tensor made in Theorem 4.1 is not original with the variant field theory. It has been made by several investigators in the past and received considerable attention by Eddington § in his affine field theory. It must be remarked, however, that the results of Eddington were not obtained by means of an axiomatic method and explicitly required invariance of the Lagrangian function. Thus, although some of the results of Eddington are similar to those obtained in this section, the principal results obtained in the remainder of this book differ markedly.

5. The Principle of Observational Covariance

The class of Lagrangian functions considered in the derivation of the variant field equations was assumed to have no preassigned transformation properties under the group of admissible coördinate transformations. Thus, although $A_{\alpha\beta}$ and $R_{\alpha\beta}$ are tensors, the equations (2.13) that they satisfy under the variant field theory are not necessarily tensorial. In the light of this noninvariance, one is naturally led to question the possible physical significance of such field equations.

On the basis of epistemological arguments in which the acceptance or

† C. W. Misner and J. A. Wheeler: "Classic Physics as Geometry," *Ann. Physics*, **2** (1957), pp. 525–603.

‡ A. S. Kompaneets: "Propagation of a Strong Electromagnetic–Gravitational Wave in Vacuo," *J. Exptl. Theoret. Phys.* (*U.S.S.R.*), **37** (1959), pp. 1722–1726 (*Soviet Physics JETP*, **37** (1960), pp. 1218–1220) points out certain doubts concerning the recovery of (4.3) by means of singularities in the field equation's solutions.

§ A. S. Eddington: *The Mathematical Theory of Relativity*, Second Edition, Cambridge University Press, London (1924).

rejection of hypothesis rests on "the observable facts of experience," Einstein stated the following principle of general covariance: "*The general laws of nature are to be expressed by equations that hold good for all systems of coordinates, that is, are covariant with respect to any* ['*coordinate*'] *substitutions whatever* (*generally covariant*)."† This principle, according to Einstein and others, implies that the equations describing fields as well as those describing matter must be tensorial.

The statement that *both* the equations describing fields and the equations describing matter must be tensorial will be found to be in error if one goes back to the epistemological criterion upon which the principle of general covariance is founded, namely, "observable facts of experience." The most that is required by this epistemological criterion is adherence to **the principle of observational covariance:** *observable quantities are to be expressed by equations that hold good for all admissible systems of coördinates.* To require adherence to anything more than the principle of observational covariance will be shown to rest either on ad hoc assumptions or on an expansion of the epistemological criterion. Requiring both the field equations and the equations that describe matter to be tensorial is easily seen to be sufficient for an adherence to the principle of observational covariance. It becomes a necessary condition, however, only if both the fields and matter are observable quantities.

Matter, per se, is a priori observable on a macroscopic level, in that it forms the basis of our known facts of experience, and we know how to make observations of macroscopic matter in the laboratory. Such observations are represented mathematically by describing the fibers of the matter continuum (see Secs. 2 and 3 of Chap. VI for a discussion of the fibers of the matter continuum). When we say that we observe fields, however, what we mean is that we observe matter being acted upon by something that we term a field. We do not, in fact, observe the fields themselves, but only their effects on "test particles." Our basic meaning of *field*—the way in which we fundamentally define a field—is given in terms of interactions with test particles. When we write down systems of equations to "describe" a field, we mean that these equations describe quantities determining the magnitude, direction, and so forth, of the interactions of a something termed a field, the presence of which results in certain characteristic changes in the fibers of the matter continuum. From the "observable facts of experience" criterion, a field thus has no a priori existential properties in itself, but rather is an abstraction arrived at by a classification of the types and natures of interactions between matter and the something that we term a field. Thus, in order to satisfy the principle of observational

† A. Einstein: *op. cit.*, p. 776.

covariance, we need only to require that the equations describing matter and the interactions of fields with matter be covariantly formulated, but not that the field equations themselves have such properties. In other words, general covariance of field equations is a sufficient, but not a necessary, condition for adherence to the principle of observational covariance. A necessary and sufficient condition is that the field-matter interactions and the matter continuum be described by covariantly formulated equations.

6. Representation of Laboratory Data

In view of the generality incorporated in the axioms of field space, particularly with respect to the lack of any assumed metric structure, the following question naturally arises: How can such an abstract, noninvariant theory have any relation to the concrete data measured in the laboratory, even if one requires adherence to the principle of observational covariance? (A related question arises in the Einstein relativity theory when one attempts to state initial data for the problem of two bodies. The difficulty there lies in the fact that although relativity theory is cast in a well-defined metric space, the metric tensor is known only after the gravitational field equations have been solved. Hence it is not clear how to specify initial data for two different points when space-time relations and structure are determined only after the field equations have been solved using this required initial data.)

The answer to our question requires an examination of the use of laboratory data and recourse to the principle of observational covariance. Granted that there is an underlying 4-dimensional ground space with structure defined by the axioms of field space, the fields defined over such a space are determined by solutions of the variant field equations with appropriate initial and boundary data. For purposes of discussion, let us divide laboratory data into two classes: first-class data that establish initial and boundary conditions required in obtaining solutions of the field equations, and second-class data that are used in checking the validity and predictions of the solutions obtained.

Laboratory data are, a priori, observable data, and hence are representable in a covariant manner in field space under the principle of observational covariance. Suppose, for the sake of simplicity, that we are concerned with a vector quantity, such as a velocity or a force, which must be determined as initial data. (The following argument goes through equally well for a tensor or collection of tensors of arbitrary rank, but then it is more complicated in its bookkeeping and tends to cloud the issue.) Since the field equations have not yet been solved, in that we do not have the required initial and boundary data, we have no knowledge of the particular underlying geometric structure and fields at this point. Not having such knowledge does not trouble us in the least in measuring such a vector quantity. We *choose*, at will, a convenient coördinate system in ordinary three space with the point of application of the vector as the origin, and obtain three numbers by decomposing the vector with respect to this coördinate system. Recording the time at which we made the measurement, we arrive at an array of four numbers, say σ^i, which may be represented as a vector in a Lorentz space. Our *choice* of the laboratory coördinate system is arbitrary to within a transformation of the homogeneous Lorentz group and has nothing whatsoever to do with the structure or coördinate system of an abstract underlying geometry. We must therefore be able to perform such homogeneous Lorentz transformations of the laboratory coördinate system in a manner consistent with the structure of field space without affecting in any way the structure or coördinate systems of this space. Otherwise, we would be led to the absurdity that the hypothesized underlying space–time geometry can be affected by the way we happen to choose our laboratory coördinate system.

Since the underlying space is assumed to be a field space, its geometry and structure are determined by solutions of the variant field equations with appropriate initial and boundary data. Not knowing the initial and boundary data at the time when we make experiments of the first class, we do not know the geometry and structure of the ground space in any extent greater than that determined by the axioms of field space. These axioms tell us, however, that the underlying space has an affine connection, a curvature tensor, and, most important, a coördinate covering for any connected compact set under consideration; further, this coördinate cover may be subjected to arbitrary admissible coördinate transformations. Once we have chosen our laboratory coördinate system without any recourse to an unknown underlying geometry (and we shall now assume that its choice is fixed—for example, at one minute to twelve it is the north-east bottom corner of our laboratory), the arbitrary admissible coördinate transformations of field space can have no effect on the laboratory coördinate system or on the values σ^i determined by its use. We are thus led to the following conclusion:

Any relations that can be established between laboratory data of the first class and quantities in abstract field space must exhibit the following properties:

(a) *The laboratory data must be transformable under the homogeneous Lorentz group without any resulting effect on the structure of field space.*

(b) *Arbitrary admissible coördinate transformations of field space must leave the laboratory data unchanged.*

We shall now show that there are relations exhibiting these properties and that they are *derivable* under the axioms of field space. In order to show this, we shall make extensive use of the material presented in Appendix A at the end of the book.

Under the axioms of field space, we know that there exist a skew-symmetric tensor $A_{\alpha\beta}$ and a symmetric tensor $R_{\alpha\beta}$. Following the arguments of Sections 1 and 2 of the Appendix, we know that there exist four linearly independent eigenvectors $\underset{j}{U^\alpha}$ of the tensor $A_{\alpha\beta} R^{\beta\eta}$ and a reciprocal set $\overset{j}{U}_\alpha$ provided

$$A_{\alpha\beta} A_{\delta\gamma} R^{\alpha\delta} R^{\beta\gamma} \neq 0,$$

which we shall assume to be true (see footnote † of page 184). Under the normalization given in the Appendix, these eigenvectors are determined only to within the transformations (A-1.20) and (A-1.21). (Equation numbers preceded by the letter A refer to equations in Appendix A.) These transformations are completely independent of coördinate transformations in field space and are simply interpreted as a collection of relations between the different possible *choices* of eigenvectors.

Let B^α be a vector in field space, evaluated at a point (0). By (A-2.3) the corresponding nonholonomic components of B^α are given by

$$b^j = Z_\alpha{}^j B^\alpha, \tag{6.1}$$

where

$$Z_\alpha{}^j = \overset{j}{U}_\alpha, \tag{6.2}$$

and similarly

$$B^\alpha = Z^\alpha{}_j b^j, \tag{6.3}$$

where

$$Z^\alpha{}_j = \underset{j}{U^\alpha}. \tag{6.4}$$

Now set

$$b^j = l^j{}_k \sigma^k, \quad \sigma^j = \overset{-1}{l}{}^j{}_k b^k, \tag{6.5}$$

where $((l^j{}_k))$ is the nonsingular invariant numerical matrix given by (A-2.16). Substituting (6.5) into (6.3), we have

$$B^\alpha = Z^\alpha{}_j \, l^j{}_k \, \sigma^k, \tag{6.6}$$

and similarly, by (6.1) and (6.5),

$$\sigma^j = \overset{-1}{l}{}^j{}_k \, Z_\alpha{}^k \, B^\alpha. \tag{6.7}$$

The results leading up to and including (A-2.18) state that the σ^j defined by (6.7) transform like the components of a vector under an element of the homogeneous Lorentz group whenever we make a change in our *choice* of eigenvectors according to (A-1.20) and (A-1.21). Of more importance, it is proved that the above result holds for any B^α, and is independent of any and all admissible coördinate transformations in field space because of the invariance of the inner product $Z_\alpha{}^k B^\alpha$, and the collection of all σ's obtained from all B's by (6.7) forms a Lorentz vector space. Conversely, given any Lorentz vector σ^j, (6.6) defines a vector in field space, which is invariant for any transformation of the homogeneous Lorentz group on the σ's, and (6.6) transforms any Lorentz vector space into a vector space of vectors defined over the point (0) in field space.

From the above results we see that equations (6.6) and (6.7) exhibit the required properties of a relation between laboratory data of the first class and quantities in field space. Moreover, since the set of all σ's forms a Lorentz vector space and hence exhibits all of the structure of experimental data, there is no loss of generality in using (6.6) and (6.7) as the connecting equations, particularly as the laboratory datum σ^j gives a *vector* B^α in field space and thus satisfies the requirements of the principle of observational covariance. It should also be noted that the Lorentz structure of the laboratory data arises from our ability to *choose* different *coördinate systems* to represent the results of our experiments, and that the proven Lorentz structure of (6.6) and (6.7) arises from our ability to *choose* various *systems of eigenvectors* to within the relations (A-1.20) and (A-1.21). Equations (6.6) and (6.7) are thus acceptable not only mathematically but also because *they exactly reflect the freedom that can be exercised by the investigator in his choice of coördinate systems.*

There is a very important point to be noted concerning the connecting equations (6.6) and (6.7) with respect to data of the first class. In most initial-value and boundary-value problems, the initial and boundary data can be specified in a manner independent of the auxiliary field or fields that are sought as solutions to the problem. This is not true here. The quantities $Z^\alpha{}_j$ and $Z_\alpha{}^j$ in (6.6) and (6.7) are eigenvectors of $A_{\alpha\beta} \, R^{\beta\eta}$ and hence depend intrinsically on the curvature fields in field space. Thus, given specific laboratory data σ^j, the B^α determined by (6.6) as initial data are fixed only to within our knowledge of the Z's. Such a situation is to be expected, however, in view of our ability to perform an arbitrary admissible coördinate transformation in field space without affecting the geometry or fields. If such structure were not exhibited by (6.6), the obtained solutions to the field equations would be intrinsically tied to our arbitrary choice in determining the laboratory coördinate system and hence not transformable under the admissible group of transformations in field space. In a similar manner, the form of (6.6) is also required because of the differences between the groups of transformations allowed in field space and in the space of laboratory measurements (that is, the arbitrary admissible group versus the homogeneous Lorentz group).

The question of second-class data now is easily answered. Suppose we have a solution to the variant field equations that has been obtained using the required initial and boundary data determined by use of (6.6) and (6.7). We then know $A_{\alpha\beta}$, $R_{\alpha\beta}$ and the eigenvectors everywhere in the domain under consideration, and we also know B^{α}. All terms on the right-hand side of (6.7) are thus known, and hence we would predict a field of σ's that can be compared with experimental data of the second class since the σ's thereby obtained have all the structure of experimental data (for example, σ^{j} is a Lorentz vector).

It is to be expressly noted that we have derived the connecting equations (6.6) and (6.7) under the assumption that the Maxwellian field tensor $A_{\alpha\beta}$ is such that the relation

$$A_{\alpha\beta}\, A_{\delta\gamma}\, R^{\alpha\delta}\, R^{\beta\gamma} \neq 0$$

holds (that is, the Maxwellian field does not vanish in field space). If we do not make this assumption, nor equivalently assume that there is a skew-symmetric tensor field defined over the geometric space of definition with the properties assumed of $A_{\alpha\beta}$, we cannot derive (6.6) but rather must assume ad hoc relations between geometry, fields, and experimental data. (It is easily seen by examination of Appendix A that all the results stated there continue to hold under this alternative assumption.) It is shown in Chapters V through VIII that the tensor $A_{\alpha\beta}$ is either equal to or proportional to the electromagnetic field tensor, and hence electromagnetic phenomena are inherently necessary if relations between experimental data and field-theoretic data are to be *derived* results. It is thus evident why in the Einstein theory there is significant difficulty in attempting to formulate the boundary data when only gravitational phenomena are considered. This is also substantiated by Theorem 3.2, which states that in the Einstein gravitational theory there is no natural geometric skew-symmetric tensor to take the place of $A_{\alpha\beta}$.

It will be shown in Appendix A that to every vector B^α in field space at (0) there corresponds a spinor $\mathfrak{B}$ given by (A-3.7):

$$\mathfrak{B} = \mathfrak{G}_\alpha B^\alpha, \tag{6.8}$$

where the matrices $\mathfrak{G}_\alpha$ are linear homogeneous matrix functions of the $Z_\alpha{}^j$ given by (A-3.1). Because of the form of the $\mathfrak{G}_\alpha$, we may write

$$\mathfrak{G}_\alpha = \mathfrak{Q}_j Z_\alpha{}^j, \tag{6.9}$$

where the $\mathfrak{Q}_j$ are invariant numerical matrices. Substituting (6.9) into (6.8) and using (6.3), we have

$$\mathfrak{B} = \mathfrak{Q}_j Z_\alpha{}^j B^\alpha = \mathfrak{Q}_j b^j.$$

Hence, on using (6.5) we obtain

$$\mathfrak{B} = \mathfrak{Q}_j l^j{}_k \sigma^k. \tag{6.10}$$

Equation (6.10) states that the spinors considered in the Appendix are in actuality spinors over the space of experimental data, by the interpretations established concerning (6.6) and (6.7). As such, they are experimentally verifiable to the extent that spinors are verifiable in the laboratory. Thus the spinor algebra and *analysis* established in the Appendix may be used in exactly the same manner as in currently accepted theories without our having to consider the underlying geometric structure except through (6.6) and (6.7). The use of the spinor analysis also allows us to consider experimental data involving differentiation processes, since

$$Z^\gamma{}_i \mathfrak{D}_\gamma \mathfrak{B}$$

is well defined in the laboratory coördinate system.

From the interpretation established for equations (6.6) and (6.7), we may conclude that any tensor quantity in field space can be reduced to laboratory data by writing it in terms of its nonholonomic components. In particular, the two-blade structure of the $A_{\alpha\beta}$ field exhibited in the Appendix is obtained from the nonholonomic representations of $A_{\alpha\beta}$ and hence is verifiable, both mathematically and physically, in the laboratory.

7. Conservation Laws, Point Transformations, and Quantization

The fundamental laws of physics are embodied in statements of the conservation of basic quantities: mass, momentum, energy, charge, and so forth. The mathematical formulation of these fundamental statements involves certain integral relations of a topological, rather than a metric-differential-geometric, nature. In the theory of general relativity and also in the Einstein–Hlavatý unified field theory, conservation laws are embodied in the covariant conservation of a tensor $T^{\alpha\beta}$ of the second rank,

$$T^{\alpha\beta}{}_{;\beta} = 0.$$

To obtain the fundamental topological conservation laws from these theories, a nontensorial quantity $t^{\alpha}{}_{\beta}$ must be introduced by means of the identities satisfied by the curvature fields, which, together with the mixed tensor density $T^{\alpha}{}_{\beta}$ corresponding to $T^{\alpha\beta}$, satisfy the equations

$$\int (T^{\alpha}{}_{\beta} + t^{\alpha}{}_{\beta})_{,\alpha}\, dV = 0.$$

This is a most important example of the topological, rather than the differential-metric, structure of the mathematical laws of physics. If covariant, rather than topological, conservation laws are used to describe the intuitive concepts of physical conservation, significant problems arise for which there has as yet been no complete solution.†

The axioms of field space lead directly and simply to a system of fundamental laws of balance (generalized conservation laws) of the required topological nature since they arise from variational considerations. It can be proved‡ that to any system of partial differential equations arising from a variational statement in an $\mathcal{E}$, there corresponds a collection of *energy forms*

$$W^{\alpha}{}_{\beta} \stackrel{\text{def}}{=} P_{(\mu),\beta}\, \mathfrak{L}_{,P_{(\mu),\alpha}} - \delta^{\alpha}_{\beta}\, \mathfrak{L}, \tag{7.1}$$

where $P_{(\mu)}$ is the collection of functions varied in the variational statement, such that

$$W^{\alpha}{}_{\beta,\alpha} = -\partial_{\beta}\mathfrak{L} - \{\mathrm{E}|\mathfrak{L}\}_{P_{(\mu)}}\, P_{(\mu),\beta}. \tag{7.2}$$

Here

$$\{\mathrm{E}|\mathfrak{L}\}_{P_{(\mu)}} \stackrel{\text{def}}{=} \mathfrak{L}_{,P_{(\mu)}} - (\mathfrak{L}_{,P_{(\mu),\alpha}})_{,\alpha}$$

† For a detailed exposition of the problems associated with covariant conservation laws, see J. G. Fletcher: "Local Conservation Laws in Generally Covariant Theories," *Rev. Mod. Phys.*, **32** (1960), pp. 65–87.

‡ D. G. B. Edelen: "The Invariance Group for Hamiltonian Systems of Partial Differential Equations," *Arch. Rational Mech. Anal.*, **5** (1960), pp. 95–176.

is the Lagrangian derivative of $\mathfrak{L}$ with respect to $P_{(\mu)}$. The notation $\partial_\beta \mathfrak{L}$ in (7.2) is used to denote the partial derivative of $\mathfrak{L}$ with respect to the independent variables x^β evaluated for constant $P_{(\mu)}$. Since field space is an $\mathcal{E}$ by axiom F_1, we see that under F_1 and F_4 the above result is directly applicable in $\mathfrak{F}$. Noting that the collection of functions $P_{(\mu)}$ corresponds to the functions $L^\alpha{}_{\beta\gamma}$ and $Q_{(\rho)}$ under F_4, we have

$$W^\alpha{}_\beta = L^\eta{}_{\mu\pi,\beta}\, \mathfrak{L}_{,L^\eta{}_{\mu\pi,\alpha}} + Q_{(\rho),\beta}\, \mathfrak{L}_{,Q_{(\rho),\alpha}} - \delta^\alpha_\beta\, \mathfrak{L}. \tag{7.3}$$

Performing the indicated operations and remembering that

$$H^{\alpha\beta} = \mathfrak{L}_{,A_{\alpha\beta}}, \quad I^{\alpha\beta} = \mathfrak{L}_{,R_{\alpha\beta}}, \quad W^{(\rho)\eta} = \mathfrak{L}_{,Q_{(\rho),\eta}},$$

we obtain

$$W^\alpha{}_\beta = L^\eta{}_{\eta\mu,\beta}\,(2H^{\alpha\mu} + I^{\alpha\mu}) + Q_{(\rho),\beta}\, W^{(\rho)\alpha} - L^\alpha{}_{\eta\mu,\beta}\, I^{\eta\mu} - \delta^\alpha_\beta\, \mathfrak{L} \tag{7.4}$$

as an explicit expression for $W^\alpha{}_\beta$ in field space. *Equations* (3.6) *and* (7.2) *constitute the basic topological laws of balance in field space.* A useful mnemonic form of $W^\alpha{}_\beta$ is

$$W^\alpha{}_\beta = 2(-H^{\mu\alpha}\, L^\rho{}_{\rho\mu,\beta} + I^{\mu[\alpha}\, L^{\rho]}{}_{\rho\mu,\beta}) + W^{(\rho)\alpha}\, Q_{(\rho),\beta} - \delta^\alpha_\beta\, \mathfrak{L}, \tag{7.5}$$

which is easily obtained from (7.4).

In addition to expressing the laws of balance (7.2), the energy forms $W^\alpha{}_\beta$ are of central importance in the study of point transformations in $\mathfrak{F}$. Let $P_{(\mu)}(\mathbf{x})$ be the functions that are varied under axiom F_4, that is, let

$$P_{(\mu)} = (L^\alpha{}_{\beta\gamma}, Q_{(\rho)}),$$

and set

$$\tilde{x}^\alpha = x^\alpha + \Delta x^\alpha(\mathbf{x}) + \text{☆}, \tag{7.6}$$

$$\tilde{P}_{(\mu)}(\tilde{\mathbf{x}}) = P_{(\mu)}(\mathbf{x}) + \Delta P_{(\mu)}(\mathbf{x}) + \text{☆}, \tag{7.7}$$

where Δx^α and $\Delta P_{(\mu)}$ are collections of infinitesimal functions and ☆ stands for terms of second and higher orders. The infinitesimal representation of the variation process used in Section 2 to derive the variant field equations is given by

$$\tilde{P}_{(\mu)}(\mathbf{x}) = P_{(\mu)}(\mathbf{x}) + \delta P_{(\mu)}(\mathbf{x}) + \text{☆}, \tag{7.8}$$

from which it is evident that the system given by equations (7.6) and (7.7) represents a more general process than that considered in Section 2. We can, however, relate the process (7.7) and (7.8) in such a way that the process (7.6) and (7.7) is consistent with the process (7.8). Expanding the left-hand side of (7.7) by use of (7.6) gives

$$\tilde{\tilde{P}}_{(\mu)}(\mathbf{x}) + \tilde{\tilde{P}}_{(\mu)}(\mathbf{x})_{,\alpha}\,\Delta x^\alpha + ☆ = \tilde{P}_{(\mu)}(\tilde{\mathbf{x}})$$

$$= P_{(\mu)}(\mathbf{x}) + \Delta P_{(\mu)}(\mathbf{x}) + ☆.$$

By (7.8), we have

$$\tilde{\tilde{P}}_{(\mu)}(\mathbf{x})_{,\alpha}\,\Delta x^\alpha = P_{(\mu)}(\mathbf{x})_{,\alpha}\,\Delta x^\alpha + ☆,$$

and hence

$$\tilde{\tilde{P}}_{(\mu)}(\mathbf{x}) = P_{(\mu)}(\mathbf{x}) + \Delta P_{(\mu)}(\mathbf{x}) - P_{(\mu)}(\mathbf{x})_{,\alpha}\,\Delta x^\alpha + ☆.$$

Comparing this result with (7.8) gives the following consistency relation between Δx^α, $\Delta P_{(\mu)}$, and $\delta P_{(\mu)}$:

$$\delta P_{(\mu)}(\mathbf{x}) = \Delta P_{(\mu)}(\mathbf{x}) - P_{(\mu)}(\mathbf{x})_{,\alpha}\,\Delta x^\alpha(\mathbf{x}) + ☆. \tag{7.9}$$

Computing the variation of the functional $\mathcal{I}$ of axiom F_4 under (7.6) and (7.7) gives

$$\Delta\mathcal{I} = \int_{D^*} (\Delta\mathfrak{L})\,dV + \int_{D^*} \mathfrak{L}\,\Delta(dV).$$

Since

$$\Delta\,dV = dV(\tilde{\mathbf{x}}) - dV(\mathbf{x}) = dV(\mathbf{x})\,(|J| - 1),$$

where $|J|$ is the determinant of the Jacobian matrix of (7.6), and since

$$|J| = 1 + (\Delta x^\alpha)_{,\alpha} + ☆,$$

we obtain

$$\Delta\,dV = (\Delta x^\alpha)_{,\alpha}\,dV + ☆.$$

Similarly, we get

$$\begin{aligned}\Delta\mathfrak{L} &= \tilde{\mathfrak{L}}(\tilde{\mathbf{x}}) - \mathfrak{L}(\mathbf{x}) = \tilde{\mathfrak{L}}(\mathbf{x}) + \tilde{\mathfrak{L}}(\mathbf{x})_{,\sigma}\,\Delta x^\sigma - \mathfrak{L}(\mathbf{x}) + \star \\ &= \tilde{\mathfrak{L}}(\mathbf{x}) - \mathfrak{L}(\mathbf{x}) + \{\mathfrak{L}(\mathbf{x})_{,\sigma} + (\delta\mathfrak{L}(\mathbf{x})_{,\sigma} + \star)\}\,\Delta x^\sigma + \star \\ &= (\tilde{\mathfrak{L}}(\mathbf{x}) - \mathfrak{L}(\mathbf{x})) + \mathfrak{L}(\mathbf{x})_{,\sigma}\,\Delta x^\sigma + \star \\ &= (\mathfrak{L}_{,P_{(\mu)}}\,\delta P_{(\mu)} + \mathfrak{L}_{,P_{(\mu),\beta}}\,\delta P_{(\mu),\beta}) + \mathfrak{L}_{,\sigma}\,\Delta x^\sigma + \star.\end{aligned}$$

Thus we have

$$\Delta\mathcal{I} = \int_{D^*} \{\mathfrak{L}_{,P_{(\mu)}} - (\mathfrak{L}_{,P_{(\mu),\beta}})_{,\beta}\}\,\delta P_{(\mu)}\,dV + \int_{D^*\ominus D} \{\mathfrak{L}\,\Delta x^\alpha + \mathfrak{L}_{,P_{(\mu),\alpha}}\,\delta P_{(\mu)}\}\,N_\alpha\,dS,$$

and hence, on substituting from (7.9) into the above surface integral, we obtain

$$\Delta\mathcal{I} = \int_{D^*} \{\mathfrak{L}_{,P_{(\mu)}} - (\mathfrak{L}_{,P_{(\mu),\beta}})_{,\beta}\}\,\delta P_{(\mu)}\,dV + \int_{D^*\ominus D} \{(\mathfrak{L}\,\delta^\alpha_\beta - \mathfrak{L}_{,P_{(\mu),\alpha}}\,P_{(\mu),\beta})\,\Delta x^\beta + \mathfrak{L}_{,P_{(\mu),\alpha}}\,\Delta P_{(\mu)}\}\,N_\alpha\,dS.$$

Using (7.1), we may finally write

$$\Delta\mathcal{I} = \int_{D^*} \{\mathfrak{L}_{,P_{(\mu)}} - (\mathfrak{L}_{,P_{(\mu),\beta}})_{,\beta}\}\,\delta P_{(\mu)}\,dV + \int_{D^*\ominus D} \{\mathfrak{L}_{,P_{(\mu),\alpha}}\,\Delta P_{(\mu)} - W^\alpha{}_\beta\,\Delta x^\beta\}\,N_\alpha\,dS. \tag{7.10}$$

Equation (7.10) leads to several interesting results:

(a) For $\Delta x^\alpha = 0$, by (7.9) we have

$$\Delta P_{(\mu)} = \delta P_{(\mu)}.$$

Annulling $\Delta\mathcal{I}$ ($= \delta\mathcal{I}$ in this case) and setting

$$\delta P|_{D^*\ominus D} = 0$$

in conformity with the requirements of F_4, we obtain

$$\mathfrak{L}_{,P_{(\mu)}} - (\mathfrak{L}_{,P_{(\mu),\beta}})_{,\beta} = \{\mathrm{E}|\mathfrak{L}\}_{P_{(\mu)}} = 0. \tag{7.11}$$

This, however, is what was done in Section 2; in fact, (7.11) is just the symbolic form of the variant field equations (2.13) since

$$P_{(\mu)} = (L^{\alpha}{}_{\beta\gamma}, Q_{(\rho)}).$$

Combining this result with Theorem 2.2, we have proved the following result.

THEOREM 7.1. *If the space under consideration is a field space* $\mathfrak{F}$, *then the variation induced in the functional* $\mathcal{I}$ *by* (7.6) *and* (7.7) *is given by*

$$\Delta\mathcal{I} = \int_{D^*\ominus D} \{\mathfrak{L}_{,P_{(\mu),\alpha}}\Delta P_{(\mu)} - W^{\alpha}{}_{\beta}\,\Delta x^{\beta}\}\, N_{\alpha}\, dS. \tag{7.12}$$

(**b**) Define the surface functionals Y_{α} and $\mathcal{P}^{(\mu)}$ by

$$\left.\begin{aligned} Y_{\alpha} &\stackrel{\text{def}}{=} -\int_{D^*\ominus D} W^{\beta}{}_{\alpha}\, N_{\beta}\, dS, \\ \mathcal{P}^{(\mu)} &\stackrel{\text{def}}{=} \int_{D^*\ominus D} \mathfrak{L}_{,P_{(\mu),\beta}}\, N_{\beta}\, dS, \end{aligned}\right\} \tag{7.13}$$

so that

$$\left.\begin{aligned} dY_{\alpha} &= -W^{\beta}{}_{\alpha}\, N_{\beta}\, dS, \\ d\mathcal{P}^{(\mu)} &= \mathfrak{L}_{,P_{(\mu),\beta}}\, N_{\beta}\, dS. \end{aligned}\right\} \tag{7.14}$$

In a field space by Theorem 7.1, we have

$$\Delta\mathcal{I} = \int_{D^*\ominus D} \{\Delta x^{\alpha}\, dY_{\alpha} + \Delta P_{(\mu)}\, d\mathcal{P}^{(\mu)}\}. \tag{7.15}$$

Equation (7.15) states that (x^{α}, Y^{α}) and $(P_{(\mu)}, \mathcal{P}^{(\mu)})$ can be considered as two systems of conjugate quantities on $D^* \ominus D$ in the same sense that momentum and coördinates are conjugate quantities in Hamiltonian particle mechanics.† Since the space under consideration is a field space,

† P. Weiss: "On the Hamilton-Jacobi Theory and Quantization of a Dynamical Continuum," *Proc. Roy. Soc. London*, **A169** (1938), pp. 102–119.

and hence is affinely connected, the process of path translation is well defined. Let x^α be a point on $D^* \ominus D$, and define the functions $\bar{x}^\alpha(s)$ by

$$\bar{x}^\alpha(s) = x^\alpha + \kappa^\alpha(s), \qquad x^\alpha \in D^* \ominus D, \tag{7.16}$$

where $\kappa^\alpha(s)$ is the solution of the system of equations

$$\begin{aligned} &d\kappa^\alpha/ds = W^\alpha, \qquad \kappa^\alpha(0) = 0, \\ &W^\beta\, W^\alpha{}_{;\beta} = 0, \qquad W^\alpha(0) = V^\alpha, \qquad V^\alpha N_\alpha = \pm 1. \end{aligned} \tag{7.17}$$

When applied to every point of $D^* \ominus D$, the system (7.16) and (7.17) maps $D^* \ominus D$ into a new surface $D^* \ominus D(s)$, which is unique for sufficiently small s provided $D^* \ominus D$ is sufficiently smooth that N_α varies in a continuous fashion over $D^* \ominus D$. Such a mapping of $D^* \ominus D$ will be referred to as a *normal-parallel translation*. Under normal-parallel translation of $D^* \ominus D$, the conjugate systems (x^α, Y_α) and $(P_{(\mu)}, \mathcal{P}^{(\mu)})$ become functions of s. The important point is that, since (x^α, Y_α) and $(P_{(\mu)}, \mathcal{P}^{(\mu)})$ are conjugate Hamiltonian quantities, there exists a contact transformation with parameter s that maps (x^α, Y_α) and $(P_{(\mu)}, \mathcal{P}^{(\mu)})$ on $D^* \ominus D$ into $(x^\alpha, Y_\alpha)(s)$ and $(P_{(\mu)}, \mathcal{P}^{(\mu)})(s)$ on $D^* \ominus D(s)$, respectively. We have thus proved the following result.

THEOREM 7.2. *If $\mathfrak{F}$ is a field space, there exists a contact transformation with parameter s that maps the conjugate variables (x^α, Y_α) and $(P_{(\mu)}, \mathcal{P}^{(\mu)})$ on any surface $D^* \ominus D$ into a system of conjugate variables $(x^\alpha, Y_\alpha)(s)$ and $(P_{(\mu)}, \mathcal{P}^{(\mu)})(s)$ on the surface $D^* \ominus D(s)$ obtained from $D^* \ominus D$ by normal-parallel translation.*

(c) Let the space under consideration be a field space and set

$$\Delta P_{(\mu)}|_{D^*\ominus D} = 0.$$

By Theorem 7.1, we have

$$\Delta \mathcal{I} = -\int_{D^*\ominus D} \Delta x^\beta\, W^\alpha{}_\beta\, N_\alpha\, dS, \tag{7.18}$$

and by (7.9) and the above condition we can write

$$\delta P_{(\mu)} = -P_{(\mu),\alpha}\, \Delta x^\alpha, \quad \text{on} \quad D^* \ominus D. \tag{7.19}$$

Equation (7.19) is easily interpretable; it states that the variation $\delta P_{(\mu)}$ on $D^* \ominus D$ is equal to the negative of the field $P_{(\mu)}$ dragged over Δx^α by the point transformation (7.7).† Applying the divergence theorem to (7.18), we have

$$\Delta \mathcal{I} = -\int_{D^*} (\Delta x^\beta\, W^\alpha{}_\beta)_{,\alpha}\, dV. \tag{7.20}$$

Now, if we require $\Delta \mathcal{I}$ to vanish for all possible sets D^* contained in $\mathfrak{F}$ we must have

$$(\Delta x^\beta\, W^\alpha{}_\beta)_{,\alpha} = 0. \tag{7.21}$$

Combining the above with Theorem 7.1, we obtain the following result.

THEOREM 7.3. *If* $\mathfrak{F}$ *is a field space then the variation induced in* $\mathcal{I}$ *by* (7.6) *and* (7.7) *vanishes for arbitrary* D^* *such that the total variation* $\Delta P_{(\mu)}$ *vanishes on* $D^* \ominus D$ *if and only if the quantities* Δx^α *are such that*

$$(\Delta x^\beta\, W^\alpha{}_\beta)_{,\alpha} = 0 \tag{7.22}$$

is satisfied.

What has been said is that, if we wish to include the effects of point transformations of $\mathfrak{F}$ in the variation process of axiom F_4, then the generators of such point transformations cannot be assigned arbitrarily, but rather must satisfy (7.22).

A particularly important application of Theorem 7.3 is the following. Let the coördinate system be fixed, and let Δx^α be defined by

$$\Delta x^\alpha = \epsilon^{\alpha\lambda}\, v_\lambda(x), \tag{7.23}$$

where the $\epsilon^{\mu\lambda}$ are arbitrary quantities to within the condition that in the particular coördinate system under consideration we have

$$\epsilon^{(\mu\lambda)} \equiv 0, \qquad \epsilon^{\mu\nu}{}_{,\lambda} = 0, \qquad \det(\epsilon^{\mu\lambda}) \not\equiv 0. \tag{7.24}$$

(Note that it is necessary to assume that the coördinate system is fixed, for otherwise $(7.24)_2$ would not necessarily hold in arbitrary coördinate systems since we have not required $\epsilon^{\mu\nu}$ to be covariant constant.) Substituting (7.23) into (7.22) and using (7.2) and (7.24), we have

$$0 = \epsilon^{\beta\lambda}\, (v_\lambda\, W^\alpha{}_\beta)_{,\alpha} = \epsilon^{\beta\lambda}\, (v_{[\lambda}\, W^\alpha{}_{\beta]})_{,\alpha}.$$

† K. Yano: *The Theory of Lie Derivatives and Its Application*, North-Holland Publishing Co., Amsterdam (1957), Chap. I (the $P_{(\mu)}$ being interpreted as scalars).

Hence, since $\epsilon^{\beta\lambda}$ is nonsingular and arbitrary to within the conditions (7.24), the above condition can be satisfied if and only if v_λ is such that

$$0 = (v_{[\lambda} W^\alpha{}_{\beta]})_{,\alpha} = v_{[\lambda|,\alpha|} W^\alpha{}_{\beta]} + v_{[\lambda} W^\alpha{}_{\beta],\alpha}.$$

Thus, by (7.2) and the fact that $\mathfrak{F}$ is a field space (that is, $\{\mathcal{E}|\mathfrak{L}\}_{P_{(\mu)}} = 0$), we have

$$v_{[\lambda|,\alpha|} W^\alpha{}_{\beta]} = v_{[\lambda} \partial_{\beta]}\mathfrak{L}.$$

Now, given any solution to this equation in the fixed coördinate system under consideration, say $\bar{v}_\gamma$, the functions

$$M^\alpha{}_{\mu\lambda} \stackrel{\text{def}}{=} W^\alpha{}_{[\mu} \bar{v}_{\lambda]} \tag{7.25}$$

satisfy the equations

$$M^\alpha{}_{\mu\lambda,\alpha} = 0, \quad M^\alpha{}_{(\mu\lambda)} = 0, \tag{7.26}$$

as is evident from (7.24) and (7.25). From (7.26), we immediately have

$$M_{\mu\lambda} \stackrel{\text{de}}{=} \int_{D^*\ominus D} M^\alpha{}_{\mu\lambda} N_\alpha \, dS = \int_{D^*} M^\alpha{}_{\mu\lambda,\alpha} \, dV = 0. \tag{7.27}$$

Hence, since $M^\alpha{}_{\mu\lambda}$ can be interpreted by (7.25) as the moment of the energy forms $W^\alpha{}_\mu$ with respect to the vector v_λ, we have the result that this moment is conserved under the solution of the variant field equations.

THEOREM 7.4. *If $\mathfrak{F}$ is a field space and the vector v_γ is a solution to the system of equations*

$$v_{[\lambda|,\alpha|} W^\alpha{}_{\beta]} = v_{[\lambda} \partial_{\beta]}\mathfrak{L} \tag{7.28}$$

in a fixed coördinate system, then in that coördinate system the moment of $W^\alpha{}_\beta$ with respect to v_γ is conserved under the solution of the variant field equations.

It is to be noted that this result together with (7.2) establishes the *analogies to the theorems of conservation of momentum-energy and field angular momentum, but have been obtained without any assumptions with respect to the transformation properties of the Lagrangian function or to the effect that the underlying space has metric structure.* The lack of metric structure is characterized by the above restriction to a particular coördinate system for obtaining v_γ, althou gh there is, in general, a v_γ for each coördinate system. We can thus define functions in field space with the properties of field momentum-energy and field angular momentum.

(d) Consider those transformations (7.6) and (7.7) that result from an r-parameter continuous group of transformations. Let the canonical parameters be π^A $(A = 1, \ldots, r)$; then we have

$$\Delta x^\beta = X^\beta{}_A \, \Delta\pi^A, \qquad \Delta P_{(\mu)} = P_{(\mu)A} \, \Delta\pi^A, \tag{7.29}$$

where

$$X^\beta{}_A = \left.\frac{\partial \tilde{x}^\beta}{\partial \pi^A}\right|_{\pi^B=0}, \tag{7.30a}$$

$$P_{(\mu)A} = \left.\frac{\partial \tilde{P}_{(\mu)}}{\partial \pi^A}\right|_{\pi^B=0} \tag{7.30b}$$

are the infinitesimal generators of the group. In field space, by Theorem 7.1 we have

$$\Delta \mathcal{I} = \int_{D^* \ominus D} \{-W^\alpha{}_\beta \, X^\beta{}_A + \mathfrak{L}_{,P_{(\mu),\alpha}} P_{(\mu)A}\} \, \Delta\pi^A \, N_\alpha \, dS.$$

Since the $\Delta\pi^A$ are constants, this leads to

$$\Delta \mathcal{I} = \int_{D^*} \{\mathfrak{L}_{,P_{(\mu),\alpha}} P_{(\mu)A} - W^\alpha{}_\beta \, X^\beta{}_A\}_{,\alpha} \, dV \, \Delta\pi^A,$$

and thus the functional derivatives of $\mathcal{I}$ with respect to the π^A are given by

$$\frac{\Delta \mathcal{I}}{\Delta \pi^A} = \int_{D^*} \{\mathfrak{L}_{,P_{(\mu),\alpha}} P_{(\mu)A} - W^\alpha{}_\beta \, X^\beta{}_A\}_{,\alpha} \, dV. \tag{7.31}$$

Hence we have the following result.

THEOREM 7.5. *If $\mathfrak{F}$ is a field space and if $X^\alpha{}_A$ are the infinitesimal generator of a continuous r-parameter group of point transformations in $\mathfrak{F}$, then $\mathcal{I}$ is invariant under the given group if and only if the $P_{(\mu)A}$ satisfy the system of equations*

$$\{\mathfrak{L}_{,P_{(\mu),\alpha}} P_{(\mu)A} - W^\alpha{}_\beta \, X^\beta{}_A\}_{,\alpha} = 0. \tag{7.32}$$

Conversely, if $\mathcal{I}$ is invariant under the r-parameter continuous group of point transformations with infinitesimal generators (7.30), *then the solutions to the variant* *field equations admit the conservation laws* (7.32), *there being one conservation law for each parameter in the group.*

To date, this theorem has been principally useful in theories based on scalar-density Lagrangian functions, since such Lagrangian functions are usually invariant under point transformations. Theorem 4.5 is of even more importance in the general case in which the Lagrangian function is assumed to have arbitrary transformation properties, although previously this has been unrecognized. Let us assume that the $X^{\alpha}{}_{A}$ are such that

$$(W^{\alpha}{}_{\beta}\, X^{\beta}{}_{A})_{,\alpha} = 0;$$

then (4.32) assumes the form

$$(\mathfrak{L}_{,P_{(\mu),\alpha}}\, P_{(\mu)A})_{,\alpha} = 0.$$

For given $\mathfrak{L}$, the system of solutions to this equation for $P_{(\mu)A}$ defines the system of infinitesimal generators of the transformations on the $P_{(\mu)}$ having the property of leaving the action $\mathcal{I}$ invariant. On the other hand, suppose we know a system of transformations of the $P_{(\mu)}$ with generators $P_{(\mu)A}$, which for physical reasons are required to leave the action invariant. Substituting these $P_{(\mu)A}$ into the last equation above results in a *simultaneous system of linear equations for the determination of the most general Lagrangian function admitting the required conservation properties.* Theorem 7.5 thus leads to a systematic method of defining the most general Lagrangian function that will result in the required invariance of the action under physically preassigned transformations in the $P_{(\mu)}$'s and the corresponding conservation laws. This result allows us to use the arguments of invariance usually encountered in the quantum field theory (gauge invariance, charge conjugation, and so on) without having to require the Lagrangian function to be a scalar density. This enlarges the class of admissible Lagrangian functions and thereby opens whole new avenues of approach to some of the more trying problems in the quantum field theory.

The properties of the continuous groups characterized by (7.29) are important also in another way. Under the currently accepted procedures of field quantization, the field functions $P_{(\mu)}$ are no longer interpreted as functions in the classic sense, but rather are viewed as operators on the state amplitude Φ. Under transformations of coördinates, it is required that the state amplitude transform according to the law

$$\bar{\Phi} = U\,\Phi, \qquad U\,U^{*} = 1, \tag{7.33}$$

where U is a unitary operator, so that the norm of the state amplitude, $\Phi^*\Phi$, remains invariant. Under infinitesimal transformations characterized by (7.29), we have

$$\bar{\Phi} = (1 + \Delta U)\, \Phi + \star, \tag{7.34}$$

where ΔU must be antihermitian in accordance with $(7.33)_2$. Thus, under (7.29) we may write

$$\Delta U = iU_A\, \Delta\pi^A, \qquad U_A = \frac{1}{i}\frac{\partial U}{\partial \pi^A}\bigg|_{\pi^B=0}, \tag{7.35}$$

The question naturally arises of the relations between the transformation properties of the operators $P_{(\mu)}$ and the transformation properties of the state amplitude. Consider the "expectation" of the transformed operator $\tilde{P}_{(\mu)}(\mathbf{x})$ in the state described by the amplitude Φ: $\Phi^*\, \tilde{P}_{(\mu)}(\mathbf{x})\, \Phi$. *In adherence to the principle of observational covariance, expectation values must be invariant,* since they are supposedly measurable. Hence we can equally well calculate this expectation by use of the operation $P_{(\mu)}(\mathbf{x})$ and the transformed state amplitude. Thus, under the principle of observational covariance, we require

$$\Phi^*\, \tilde{P}_{(\mu)}(\mathbf{x})\, \Phi = \bar{\Phi}^*\, P_{(\mu)}(\mathbf{x})\, \bar{\Phi}. \tag{7.36}$$

Using (7.33) to eliminate the $\bar{\Phi}$ from (7.36) results in

$$\tilde{P}_{(\mu)}(\mathbf{x}) = U^*\, P_{(\mu)}(\mathbf{x})\, U. \tag{7.37}$$

This equation expresses the conditions of compatibility between the transformations on the operators $P_{(\mu)}(x)$ and transformations on the state amplitude Φ, under (7.29) and (7.33), in order for (7.36) to hold. For infinitesimal transformations, on using (7.29), (7.34), and (7.37), we have

$$\begin{aligned}\tilde{P}_{(\mu)}(\mathbf{x}) &= P_{(\mu)}(\mathbf{x}) - \Delta U\, P_{(\mu)}(\mathbf{x}) + P_{(\mu)}(\mathbf{x})\, \Delta U + \star \\ &= P_{(\mu)}(\mathbf{x}) + [P_{(\mu)}(\mathbf{x}), \Delta U] + \star,\end{aligned}$$

where $[P_{(\mu)}(\mathbf{x}), \Delta U]$ is the commutator of the operators $P_{(\mu)}(x)$ and ΔU. Now, by (7.8) we have

$$\tilde{P}_{(\mu)}(\mathbf{x}) - P_{(\mu)}(\mathbf{x}) = \delta P_{(\mu)}(x),$$

and hence, by (7.9), (7.29), and (7.35),

$$P_{(\mu)A}\,\Delta\pi^A - P_{(\mu)}(\mathbf{x})_{,\alpha}\,X^{\alpha}_A\,\Delta\pi^A$$
$$= [P_{(\mu)}(\mathbf{x}), iU_A\,\Delta\pi^A].$$

We thus obtain

$$i\{P_{(\mu)A} - P_{(\mu)}(\mathbf{x})_{,\alpha}\,X^{\alpha}_A\} = [U_A, P_{(\mu)}(\mathbf{x})] \tag{7.38}$$

as the final operator conditions on the field operators $P_{(\mu)}$ and the transformation operator U. Equations (7.38) together with the commutation relations implied by the conjugate nature of the systems (x^α, Y_α) and $(P_{(\mu)}, \mathcal{P}^{(\mu)})$ established in Theorem 7.2† constitute the fundamental commutation relations of the quantum field theory. We are thus able to obtain the appropriate operator algebra, so that the operator interpretation of the field variables is well defined in field space. It is to be noted that these results have been obtained without any assumptions on invariance structure of either the field operators or the Lagrangian operator.‡ All that has been required is satisfaction of the axioms of field space, assumption of the existence of a state amplitude function, and adherence to the principle of observational covariance.

8. Epilog

The structural theory of fields put forth in this chapter has been developed from a system of four axioms (F_1, F_2, F_3, and F_4) concerning the ground space. No more has been assumed than what is believed to be a minimal structure for the ground space. Nor has it been assumed that the ground space should be a metric space or even a well-based space. Therefore the variant field equations are *independent* of any metric or basing assumptions.

It must be stated that there is no direct physical justification for the assumption of a variational statement such as the one embodied in axiom F_4. The acceptance of this axiom must ultimately rest, as must the acceptance of the other three, on whether or not variant field theory provides an acceptable descriptive form for the various fields and phenomena encountered in physics. By using only the four axioms of field

† Any conjugate system of variables, in the sense used here, satisfies the classic Poisson bracket relations of canonical variables, which become commutation expressions when the field variables are interpreted as operators. See P. Weiss: *op. cit.*

‡ The same results are obtained by a number of authors under the specific assumption of invariance structure; for instance, see N. N. Bogoliubov and D. V. Shirkov: *Introduction to the Theory of Quantized Fields*, Interscience Publishers, New York (1959).

space and the principle of observational covariance, however, a broad and interesting class of results has been obtained:

(a) A system of partial differential equations, the variant field equations, has been obtained, consisting of $40 + n$ equations for the determination of the 40 components of affine connection $L^{\alpha}{}_{\beta\gamma}$ and the n quantities $Q_{(\rho)}$.

(b) Equations of Maxwellian form have been derived directly from the four axioms of field space in a simple and direct manner. This derivation leads to the identification of the tensor $A_{\alpha\beta}$ with the Maxwellian field tensor, so that Maxwellian fields and the curvature of field space are directly related. It also leads to definite constitutive relations for the current potentials $H^{\alpha\beta}$ and for the Maxwellian current S^{α}. In addition, a conservation law for the Maxwellian current is obtained, and these conclusions hold true for any admissible choice of the Lagrangian function. Thus, to obtain Maxwellian structure, one is not faced with the lengthy and circuitous procedures required in the majority of theories put forward to date; nor is one faced with choosing a system of assumptions or ad hoc field equations, which all too often have an unduly restrictive common domain of validity. Rather, as a direct consequence of the axioms of field space, Maxwellian structure is implicit in the variant field theory. In other words, equations of Maxwellian form have been obtained without making physical assumptions; rather, these equations have been developed as natural consequences of a system of four axioms concerning a general abstract space, and hold for any admissible Lagrangian function. It has also been shown that if the assumptions of the Einstein–Hlavatý theory are followed, requiring not only that the space admit torsion but also that the torsion tensor specifically must not vanish, then one of the Maxwellian equations would be lost and, hence on an ad hoc basis, its validity or the validity of some equivalent relation must be assumed.

(c) No transformation requirements have been placed on the Lagrangian function, and hence the variant field equations are not necessarily tensorial equations. However, even if the variant field equations are non-tensorial, $A_{\alpha\beta}$ and $R_{\alpha\beta}$ are still tensors, and hence the Maxwellian field is always represented by a tensor, namely $A_{\alpha\beta}$.

(d) Equations of balance of the required topological nature are a direct result of the axioms of field space. Thus the problems inherent in covariant conservation statements are not directly present in the variant field theory. In addition, the conservation considerations, together with the principle of observational covariance, lead in a direct and simple manner to the results upon which the quantum-field formalism rests, and they define a natural group of point transformations, leaving the basic conservation statements unaltered.

(e) It is shown in Appendix A that the Maxwellian field tensor $A_{\alpha\beta}$ admits a "two-blade" structure together with a basis for the construction of spinor representations of the field variables, and hence that field space admits such a structure as a direct consequence of its defining axioms. It is also shown that there are unique derivable relationships between laboratory data and quantities in field space, and that the spinor analysis given in Appendix A may be used in the same manner as in current theories.

According to these results, the prospects of constructing a field theory that presents gravitational and electrodynamics from a common geometric basis are most encouraging. This seems particularly true when we note that we still have complete freedom in the choice of the Lagrangian function, provided only that it is a member of the admissible class of such functions.

Chapter IV

The Form of the Affine Connection for Fields of the Nonnull and Seminull Classes

A GENERAL structural theory of fields was presented in Chapter III, in which the field equations resulted from the system of four axioms (F_1, F_2, F_3, F_4) characterizing field space. These axioms require field space to be a torsion-free, affinely connected, 4-dimensional Hausdorff space over which a Lagrangian function is defined and that the integral of this function over any compact connected set be stationary. The fundamental field variables involved in the stationarization, namely those quantities that are functionally varied to obtain the field equations, were the components of the affine connection $L^{\alpha}{}_{\beta\gamma}$ and the quantities $Q_{(\rho)}$, where (ρ) assumes the values $1, \ldots, n$. There are thus $40 + n$ resulting field equations for the determination of the $40 + n$ variables $L^{\alpha}{}_{\beta\gamma}$ and $Q_{(\rho)}$. Our purpose in this chapter is to examine these field equations, for arbitrary admissible Lagrangian functions, to determine certain possible representations for the components of affine connection. The results will allow us to study a large class of particular Lagrangian functions and to establish the basis for many of the analyses undertaken in Part Two.

1. Statement of the Problem

The variant field equations derived in Chapter III from the axioms F_1, F_2, F_3, F_4 are

$$H^{\alpha\epsilon}{}_{,\epsilon} = S^{\alpha}, \tag{1.1}$$

$$I^{\alpha\beta}{}_{:\gamma} = \tfrac{2}{5}\,\delta^{(\alpha}_{\gamma}\,I^{\beta)\epsilon}{}_{:\epsilon} + P^{\alpha\beta}_{\cdot\cdot\gamma}, \tag{1.2}$$

and

$$Z^{(\rho)} - W^{(\rho)\eta}{}_{,\eta} = 0, \tag{1.3}$$

where

$$H^{\alpha\beta} = \mathfrak{L}_{,A_{\alpha\beta}}, \qquad Z^{(\rho)} = \mathfrak{L}_{,Q_{(\rho)}}, \tag{1.4}$$

$$I^{\alpha\beta} = \mathfrak{L}_{,R_{\alpha\beta}}, \qquad W^{(\rho)\eta} = \mathfrak{L}_{,Q_{(\rho),\eta}}, \tag{1.5}$$

$$S^{\alpha} = \tfrac{1}{10}\,(3I^{\alpha\epsilon}{}_{:\epsilon} - 2\partial_{L^{\epsilon}{}_{\epsilon\alpha}}\mathfrak{L}), \tag{1.6}$$

$$P^{\alpha\beta}_{\cdot\cdot\gamma} = \tfrac{2}{5}\,\delta^{(\alpha}_{\gamma}\,\partial_{L^{\epsilon}{}_{\beta)\epsilon}}\mathfrak{L} - \partial_{L^{\gamma}{}_{\alpha\beta}}\mathfrak{L}, \tag{1.7}$$

and

$$I^{\alpha\beta}{}_{:\gamma} \overset{\text{def}}{=} I^{\alpha\beta}{}_{,\gamma} + L^{\alpha}{}_{\rho\gamma}\,I^{\rho\beta} + L^{\beta}{}_{\rho\gamma}\,I^{\alpha\rho} - L^{\rho}{}_{\rho\gamma}\,I^{\alpha\beta}. \tag{1.8}$$

From these and (III-2.10), we obtain the subsidiary equations

$$S^{\alpha}{}_{,\alpha} = 0, \tag{1.9}$$

$$A_{(\alpha\beta)} \equiv 0, \tag{1.10}$$

$$A_{[\alpha\beta,\gamma]} = 0, \tag{1.11}$$

$$R_{[\alpha\beta]} \equiv 0, \tag{1.12}$$

$$H^{(\alpha\beta)} \equiv 0, \tag{1.13}$$

$$I^{[\alpha\beta]} \equiv 0, \tag{1.14}$$

$$P^{\alpha\beta}_{\cdot\cdot\beta} = P^{\alpha\beta}_{\cdot\cdot\alpha} \equiv 0, \tag{1.15}$$

and

$$P^{[\alpha\beta]}_{\cdot\cdot\gamma} \equiv 0. \tag{1.16}$$

The problem to be solved is as follows: *What is the most general form that* $L^{\alpha}{}_{\beta\gamma}$ *can assume under the variant field equations* (1.1), (1.2), *and* (1.3)?

2. An Equivalence Theorem

The form of equations (1.2) does not lead to a straightforward or direct answer to the problem stated in the previous section, except in certain singular cases. For this reason we develop a system of equations that is equivalent to the system (1.2) and that leads to a direct answer to the problem.

We first examine to what extent the system (1.2) determines $L^{\alpha}{}_{\beta\gamma}$.

LEMMA 2.1. *There are at most 36 independent equations in the system* (1.2) *if* $I^{\alpha\beta}{}_{:\beta} \neq 0$.

Proof. Set

$$U_\gamma{}^{\alpha\beta} = I^{\alpha\beta}{}_{:\gamma} - \tfrac{2}{5}\,\delta^{(\alpha}_{\gamma} I^{\beta)\eta}{}_{:\eta} - P^{\alpha\beta}{}_{..\gamma}; \tag{2.1}$$

then (1.2) is equivalent to

$$U_\gamma{}^{\alpha\beta} = 0.$$

In addition, by (1.14) and (1.16) we have

$$U_\gamma{}^{[\alpha\beta]} \equiv 0$$

even if

$$U_\gamma{}^{\alpha\beta} \neq 0.$$

Thus, since field space is a 4-dimensional space by axiom F_1 and postulate E_1, there are at most 40 independent equations in the set (1.2). Contracting (1.2) with respect to either (α, γ) or (β, γ), which is equivalent to contracting (2.1) over the same indices, we have

$$U_\beta{}^{\alpha\beta} = U_\beta{}^{\beta\alpha} \stackrel{\mathrm{d}}{=} V^\alpha. \tag{2.3}$$

If

$$I^{\alpha\beta}{}_{:\beta} \neq 0,$$

then V^α vanishes identically by (2.1) and (1.15). Hence there exist 4 additional relations among the arguments of (1.2), so that there are at most 36 independent equations as claimed.

From Lemma 2.1 we conclude that in those cases in which

$$I^{\alpha\eta}{}_{:\eta} \neq 0,$$

the field equations (1.2) determine the components of affine connection only to within 4 arbitrary functions. This is evident, since if there were fewer than 4 arbitrary functions, there would be insufficient generality in the 40 resulting $L^\alpha{}_{\beta\gamma}$ to satisfy the identity $V^\alpha = 0$.

We must now distinguish between certain classes of fields.

DEFINITION 2.1. *Let* $I = \det(I^{\alpha\beta})$. *A field* $I^{\alpha\beta}$ *is said to be a member of*
(*a*) *the nonnull class if and only if* $I \neq 0$, $I^{\alpha\eta}{}_{:\eta} \neq 0$;
(*b*) *the seminull class if and only if* $I \neq 0$, $I^{\alpha\eta}{}_{:\eta} = 0$;
(*c*) *the proper null class if and only if* $I^{\alpha\beta} = 0$ *for all* (α, β); *and*
(*d*) *the improper null class if and only if* $I = 0$ *and* $I^{\alpha\beta} \neq 0$ *for some* (α, β).

We may now restate Lemma 2.1:

LEMMA 2.1′. *If* $I^{\alpha\beta}$ *is a member of the nonnull class, then there are only 36 independent equations in the system* (1.2).

In order to establish the required equivalence theorem, we shall need the following result.

LEMMA 2.2. *If* $I^{\alpha\beta}$ *is a member of either the nonnull or seminull class, then the system* (1.2) *is equivalent to the system*

$$I_{\alpha\beta:\gamma} = \tfrac{2}{5} I^{\mu\rho} I_{\gamma(\alpha} I_{\beta)\mu:\rho} - P_{\dot{\alpha}\dot{\beta}\gamma}, \tag{2.4}$$

where

$$P_{\dot{\alpha}\dot{\beta}\gamma} = P^{\mu\eta}{}_{\cdot\cdot\gamma} I_{\mu\alpha} I_{\eta\beta}, \tag{2.5}$$

and where $I_{\alpha\beta}$ *is the unique reciprocal of* $I^{\alpha\beta}$, *defined by*

$$I\, I_{\alpha\beta} = I_{,I^{\alpha\beta}}. \tag{2.6}$$

Proof. By (1.14), the field $I^{\alpha\beta}$ is symmetric. Since $I^{\alpha\beta}$ is assumed to be a member of either the nonnull or the seminull class, the quantity I is unequal to zero, and hence there exists a unique reciprocal field $I_{\alpha\beta}$, defined by (2.6), with the property

$$I^{\alpha\beta} I_{\beta\gamma} = \delta^{\alpha}_{\gamma}. \tag{2.7}$$

Thus, upon multiplying (1.2) by $I_{\alpha\eta} I_{\beta\mu}$, noting that

$$I^{\alpha\beta}{}_{:\gamma} I_{\alpha\eta} I_{\beta\mu} = -I_{\eta\mu:\gamma},$$

which immediately follows from (2.7), and relabeling the free indices, we obtain (2.4). The converse is easily established by using $I^{\alpha\beta}$ to pull the indices in (2.4) and thus to obtain (1.2).

THEOREM 2.1: *If* $I^{\alpha\beta}$ *is a member of either the nonnull or the seminull class, then the system* (1.2) *is equivalent to the system*

$$I_{\alpha\beta:\gamma} = -C_{\gamma\alpha\beta}, \tag{2.8}$$

where

$$C_{\gamma\alpha\beta} = 2I_{\gamma(\alpha} M_{\beta)} + P_{\dot{\alpha}\dot{\beta}\gamma} \tag{2.9}$$

for arbitrary M_{α}.

Proof. Since $I^{\alpha\beta}$ is assumed to be a member of either the nonnull class or the seminull class, we may use Lemma 2.2 to replace the system (1.2) by the equivalent system (2.4). Eliminating the colon derivative in (2.4) by (2.8) gives

$$C_{\gamma\alpha\beta} = \tfrac{2}{5} I^{\mu\rho} I_{\gamma(\alpha} C_{|\rho|\beta)\mu} + P_{\dot{\alpha}\dot{\beta}\gamma}. \tag{2.10}$$

Consider first the case in which $P_{\dot{\alpha}\dot{\beta}\gamma} = 0$. We then have, in place of (2.10),

$$C_{\gamma\alpha\beta} = \tfrac{1}{5} (C^{\mu}_{\cdot\beta\mu} I_{\gamma\alpha} + C^{\mu}_{\cdot\alpha\mu} I_{\gamma\beta}), \tag{2.11}$$

where

$$C^{\mu}_{\cdot\beta\mu} = I^{\mu\rho} C_{\rho\beta\mu}.$$

Set

$$C^{\eta}_{\cdot\alpha\eta} = 5M_{\alpha}, \tag{2.12}$$

where M_{α} is to be determined. Substituting (2.12) into (2.11) gives

$$C_{\gamma\alpha\beta} = 2I_{\gamma(\alpha} M_{\beta)}, \tag{2.13}$$

which is a general solution to (2.11), provided M_{α} is such that it satisfies (2.12). Substituting (2.13) into (2.12) gives an identity in M_{α}, and hence (2.13) is a general solution to (2.11) for arbitrary M_{α}. Since (2.11) is the homogeneous part of (2.10) and since (2.10) is a linear equation in $C_{\gamma\alpha\beta}$, the general solution is given by the sum of a particular solution and the general homogeneous solution. Such a particular solution is given by

$$\overset{\circ}{C}_{\gamma\alpha\beta} = P_{\dot{\alpha}\dot{\beta}\gamma}.$$

To see this, we need only to note by (1.15) that

$$\overset{\circ}{C}{}^{\mu}_{\cdot\alpha\mu} = P_{\dot{\alpha}\dot{\mu}}{}^{\mu} = P^{\rho\mu}_{\cdot\cdot\mu} I_{\rho\alpha} = 0.$$

Thus the general solution to (2.10) is given by (2.9).

THEOREM 2.1′. *If $I^{\alpha\beta}$ is a member of either the nonnull or the seminull class, then the system* (1.2) *is equivalent to the system*

$$I^{\alpha\beta}{}_{:\gamma} = 2\delta^{(\alpha}_{\gamma} I^{\beta)\eta} M_{\eta} + P^{\alpha\beta}_{\cdot\cdot\gamma} \tag{2.14}$$

for arbitrary M_{η}.

Proof. The result follows immediately from Theorem 2.1 and the identity

$$I^{\alpha\beta}{}_{:\gamma} = -I^{\alpha\sigma} I^{\beta\rho} I_{\sigma\rho:\gamma}.$$

COROLLARY. *If $I^{\alpha\beta}$ is a member of either the nonnull or the seminull class, then the system* (1.2) *implies*

$$I^{\alpha\eta}{}_{:\eta} = 5I^{\alpha\rho} M_\rho \tag{2.15}$$

for arbitrary M_ρ, so that

$$S^\alpha = \tfrac{3}{2} I^{\alpha\rho} M_\rho - \tfrac{1}{5} \partial_{L^\sigma{}_{\sigma\alpha}} \mathfrak{L}. \tag{2.16}$$

Proof. The result is obtained by contracting (2.14), using (1.15), and substituting the results into (1.6).

We may now distinguish between $I^{\alpha\beta}$ fields of the nonnull and the seminull class:

THEOREM 2.2. *If $I^{\alpha\beta}$ is a symmetric form with nonvanishing determinant, then the system* (1.2) *is equivalent to the system*

$$I_{\alpha\beta:\gamma} = -C_{\gamma\alpha\beta}, \tag{2.17}$$

with

$$C_{\gamma\alpha\beta} = 2I_{\gamma(\alpha} M_{\beta)} + P_{\dot{\alpha}\dot{\beta}\gamma}. \tag{2.18}$$

Further, $I^{\alpha\beta}$ is a member of the nonnull class if $M_\alpha \neq 0$, and of the seminull class if $M_\alpha = 0$.

Proof. Since $I^{\alpha\beta}$ has a nonzero determinant by hypothesis, $I^{\alpha\beta}$ is a member of either the nonnull or the seminull class. Theorem 2.1 then holds and thus establishes the equivalence of the system (1.2) and the equations (2.17) and (2.18). By the corollary to Theorem 2.1′, we have

$$I^{\alpha\eta}{}_{:\eta} = 5I^{\alpha\sigma} M_\sigma.$$

Thus, since $I^{\alpha\beta}$ is a nonsingular symmetric form by hypothesis, $I^{\alpha\eta}{}_{:\eta}$ will or will not vanish accordingly as M_σ vanishes or does not vanish. Hence, $I^{\alpha\beta}$ is a member of the nonnull class (the seminull class) if $M_\alpha \neq 0$ ($M_\alpha = 0$).

Theorem 2.2 allows us to obtain results for nonnull and seminull class fields simultaneously. We need only to require that $I^{\alpha\beta}$ have a nonvanishing determinant and that either M_α be nonzero or vanish in order to obtain results for nonnull and seminull fields, respectively.

It is of interest to note the similarity between equations (2.17) and the postulate B_3 of a well-based space. The similarity is even more evident since the remaining postulates B_1 and B_2 are satisfied, to within the condition that $I^{\alpha\beta}$ be a tensor field, by the assumption that $I^{\alpha\beta}$ is a member of the nonnull or seminull class, and the fact that

$$I^{[\alpha\beta]} = 0$$

by definition. Because of this formal similarity, we shall see that many of the results in a well-based space hold in a field space with nonnull or seminull $I^{\alpha\beta}$ fields.

3. The General Form of the Affine Connection for Fields of the Nonnull and Seminull Classes

The equivalence Theorems 2.1 and 2.2 provide a direct and straightforward means of answering the question posed in Section 1 for fields of the nonnull and seminull classes. By these theorems, the system (1.2) may be replaced by the equivalent system

$$I_{\alpha\beta:\gamma} = -2I_{\gamma(\alpha}\, M_{\beta)} - P_{\dot\alpha\dot\beta\gamma} \tag{3.1}$$

for arbitrary M_α. Expanding the colon derivative in (3.1) by means of (1.8) and noting that

$$I_{\alpha\beta:\gamma} = -I_{\alpha\sigma}\, I_{\beta\rho}\, I^{\sigma\rho}{}_{:\gamma},$$

we obtain

$$I_{\alpha\beta,\gamma} - L^\sigma{}_{\beta\gamma}\, I_{\alpha\sigma} - L^\sigma{}_{\alpha\gamma}\, I_{\sigma\beta}$$

$$= -2I_{\gamma(\alpha}\, M_{\beta)} - L^\sigma{}_{\sigma\gamma}\, I_{\alpha\beta} - P_{\dot\alpha\dot\beta\gamma} \overset{\text{def}}{=} -W_{\gamma\alpha\beta}. \tag{3.2}$$

Interpreting (3.2) as an algebraic equation for the determination of the $L^\alpha{}_{\beta\gamma}$, by a direct application of Christoffel's eliminative method we obtain

$$L^\alpha{}_{\beta\gamma} = \Gamma^\alpha{}_{\beta\gamma}(I_{\eta\pi}) + \tfrac{1}{2}\, I^{\alpha\rho}\, (W_{\beta\rho\gamma} - W_{\rho\gamma\beta} + W_{\gamma\beta\rho}), \tag{3.3}$$

where the $\Gamma^\alpha{}_{\beta\gamma}(I_{\eta\pi})$ are the Christoffel symbols of the second kind formed from $I_{\eta\pi}$:

$$\Gamma^\alpha{}_{\beta\gamma}(I_{\eta\pi}) = \tfrac{1}{2}\, I^{\alpha\rho}\, (I_{\rho\gamma,\beta} - I_{\gamma\beta,\rho} + I_{\beta\rho,\gamma}). \tag{3.4}$$

This is not a complete solution for the affine connection, however, since by $(3.2)_2$ the $W_{\alpha\beta\gamma}$ in (3.3) are functions of $L^{\sigma}{}_{\sigma\gamma}$. Expanding $W_{\gamma\alpha\beta}$ in (3.3) by means of $(3.2)_2$ yields

$$\tfrac{1}{2} I^{\alpha\sigma} (W_{\beta\sigma\gamma} - W_{\sigma\gamma\beta} + W_{\gamma\beta\sigma}) = M_\sigma I^{\sigma\alpha} I_{\beta\gamma} + \tfrac{1}{2} (L^{\rho}{}_{\rho\beta}\, \delta^{\alpha}_{\gamma} + L^{\rho}{}_{\rho\gamma}\, \delta^{\alpha}_{\beta} - L^{\rho}{}_{\rho\sigma} I^{\sigma\alpha} I_{\beta\gamma}) + \tfrac{1}{2} (P_{\dot{\beta}}{}^{\alpha.}{}_{.\dot{\gamma}} - P^{\alpha}{}_{.\dot{\gamma}\dot{\beta}} + P_{\dot{\gamma}\dot{\beta}}{}^{\alpha}{}_{.}). \tag{3.5}$$

Thus, substituting (3.5) into (3.3) and contracting gives

$$L^{\alpha}{}_{\alpha\gamma} = \Gamma^{\alpha}{}_{\alpha\gamma}(I_{\eta\pi}) + M_\gamma + 2L^{\alpha}{}_{\alpha\gamma} + \tfrac{1}{2} (P_{\dot{\alpha}}{}^{\alpha.}{}_{.\dot{\gamma}} - P^{\alpha..}{}_{.\gamma\dot{\alpha}} + P_{\dot{\gamma}\dot{\alpha}}{}^{\alpha}{}_{.}).$$

Since, by (1.15),

$$P^{\alpha..}{}_{.\gamma\alpha} = P_{\gamma\alpha}{}^{\alpha}{}_{.} = 0,$$

we thus obtain

$$L^{\alpha}{}_{\alpha\gamma} = \Gamma^{\alpha}{}_{\alpha\gamma}(I_{\eta\pi}) + M_\gamma + 2L^{\alpha}{}_{\alpha\gamma} + \tfrac{1}{2} P_{\dot{\alpha}}{}^{\alpha.}{}_{.\dot{\gamma}}.$$

This may be solved for $L^{\alpha}{}_{\alpha\gamma}$ in terms of the other variables to yield

$$L^{\alpha}{}_{\alpha\gamma} = -\Gamma^{\alpha}{}_{\alpha\gamma}(I_{\eta\pi}) - M_\gamma - \tfrac{1}{2} P_{\dot{\alpha}}{}^{\alpha.}{}_{.\dot{\gamma}}. \tag{3.6}$$

Finally, substituting (3.6) into the right-hand side of (3.5) and then substituting this result into (3.3), we get

$$L^{\alpha}{}_{\beta\gamma} = \Gamma^{\alpha}{}_{\beta\gamma}(I_{\eta\pi}) - \tfrac{1}{2} (M_\gamma\, \delta^{\alpha}_{\beta} + M_\beta\, \delta^{\alpha}_{\gamma} - 3M_\sigma I^{\sigma\alpha} I_{\beta\gamma}) - \tfrac{1}{2} (\Gamma^{\rho}{}_{\rho\beta}(I_{\eta\pi})\, \delta^{\alpha}_{\gamma} + \Gamma^{\rho}{}_{\rho\gamma}(I_{\eta\pi})\, \delta^{\alpha}_{\beta} - \Gamma^{\rho}{}_{\rho\sigma}(I_{\eta\pi}) I^{\sigma\alpha} I_{\beta\gamma}) + \tfrac{1}{2} (P_{\dot{\beta}}{}^{\alpha.}{}_{.\dot{\gamma}} - P^{\alpha}{}_{.\dot{\gamma}\dot{\beta}} + P_{\dot{\gamma}\dot{\beta}}{}^{\alpha}{}_{.}) - \tfrac{1}{4} (P_{\dot{\sigma}}{}^{\sigma.}{}_{.\dot{\beta}}\, \delta^{\alpha}_{\gamma} + P_{\dot{\sigma}}{}^{\sigma.}{}_{.\dot{\gamma}}\, \delta^{\alpha}_{\beta} - P_{\dot{\sigma}}{}^{\sigma\alpha}{}_{..} I_{\beta\gamma}) \tag{3.7}$$

as the general form of the affinity in the case of $I^{\alpha\beta}$ fields of the nonnull and seminull classes. Equation (3.7) gives $L^{\alpha}{}_{\beta\gamma}$ as an explicit function of the 50 functions $I^{\alpha\beta}$, M_α, and $P^{\alpha\beta}{}_{..\gamma}$. (It follows from (2.15) and (2.16) that there are only 36 independent components of $P^{\alpha\beta}{}_{..\gamma}$.) If we assign these 50 functions arbitrarily and then substitute (3.7) into the system (1.2), the system reduces to an algebraic identity. In this sense, not only is (3.7) a general solution for the affinity from (1.2), but in addition the functions $I^{\alpha\beta}$, M_α, and $P^{\alpha\beta}{}_{..\gamma}$ may be interpreted as potential functions for the affinity.

It now remains to show that (3.7), together with the variant field equations, yields a deterministic system of equations for the functions $I^{\alpha\beta}$, M_α, $P^{\alpha\beta\cdot}_{\cdot\cdot\gamma}$, and $Q_{(\rho)}$, the former three sets being interpreted as potential functions for the affinity. Examining the field equations (1.1)–(1.3) together with the subsidiary defining equations (1.4)–(1.7), we see that (1.2) may be eliminated since it is identically satisfied by (3.7) in terms of the potential functions $I^{\alpha\beta}$, M_α, and $P^{\alpha\beta\cdot}_{\cdot\cdot\gamma}$. We must, however, adjoin to the remaining field equations (1.1) and (1.3) the equations (1.5) and (1.7), so that the fields $I^{\alpha\beta}$ and $P^{\alpha\beta\cdot}_{\cdot\cdot\gamma}$ in (3.7) are related to the derivatives of the Lagrangian function in a manner consistent with the terms in (1.2). We thus obtain the following system of $50 + n$ field equations:

$$
\begin{aligned}
&H^{\alpha\sigma}{}_{,\sigma} = S^\alpha, \\
&I^{\alpha\beta} = \mathfrak{L}_{,R_{\alpha\beta}}, \\
&P^{\alpha\beta\cdot}_{\cdot\cdot\gamma} = \tfrac{2}{5}\,\delta^{(\alpha}_{\gamma}\,\partial_{L^\sigma{}_{\beta)\sigma}}\mathfrak{L} - \partial_{L^\gamma{}_{\alpha\beta}}\mathfrak{L}, \\
&Z^{(\rho)} - W^{(\rho)\eta}{}_{,\eta} = 0, \quad \rho = 1, \ldots, n,
\end{aligned}
\tag{3.8}
$$

for the determination of the $50 + n$ functions $I^{\alpha\beta}$, M_α, $P^{\alpha\beta\cdot}_{\cdot\cdot\gamma}$, and $Q_{(\rho)}$ under the general potential solution for the affinity given by (3.7). We have thus proved the following fundamental result.

Theorem 3.1. *If $I^{\alpha\beta}$ is a member of either the nonnull or the seminull class, then the most general form that $L^\alpha{}_{\beta\gamma}$ can assume under the axioms of field space is given by* (3.7) *in terms of the 50 potential functions $I^{\alpha\beta}$, M_α, and $P^{\alpha\beta\cdot}_{\cdot\cdot\gamma}$, which, together with the n functions $Q_{(\rho)}$, must satisfy the 50 + n equations* (3.8).

It is of interest to note that according to Theorem 3.1 the Maxwellian field tensor $A_{\alpha\beta}$ assumes the form

$$A_{\alpha\beta} = -2M_{[\beta,\alpha]} - P^{\cdot\,\sigma}_{\sigma\cdot\,[\beta,\alpha]}, \tag{3.9}$$

as is seen by substituting (3.6) into (II-2.5) and noting that

$$\Gamma^\sigma{}_{\sigma[\beta}(I_{\eta\pi})_{,\gamma]} = (\ln\sqrt{|I|})_{,[\beta\gamma]} = 0.$$

Hence $A_{\alpha\beta}$ will be a null tensor only if

$$M_{[\beta,\alpha]} = -\tfrac{1}{2}\,P^{\cdot\,\sigma}_{\sigma\cdot\,[\beta,\alpha]}. \tag{3.10}$$

The form of the affine connection that results for proper null fields and improper null fields is readily obtained in those cases in which it is required. For this reason, we leave the derivation of $L^{\alpha}{}_{\beta\gamma}$ for such fields to the sections in which they are discussed.

4. Generalized Conformal, Gauge, and Projective Transformations of the Affinity

The theory of Weyl geometry yields affinities that are invariant under simultaneous conformal and gauge transformations,†

$$'g_{\alpha\beta} = \sigma(x) g_{\alpha\beta}, \tag{4.1}$$

$$'f_{\gamma} = f_{\gamma} + 4[\ln \sigma(x)]_{,\gamma}. \tag{4.2}$$

The question naturally arises of the properties of the affinity (3.7) under transformations of a similar nature.

DEFINITION 4.1. *A field $I^{\alpha\beta}$ is said to undergo a generalized conformal transformation to an image field $'I^{\alpha\beta}$ if and only if*

$$'I^{\alpha\beta} = \sigma(x)\, I^{\alpha\beta}, \tag{4.3}$$

where the generator $\sigma(x)$ is an arbitrary function of class C^1 that is bounded away from zero.

DEFINITION 4.2. *A field M_β is said to undergo a generalized gauge transformation to an image field $'M_\beta$ if and only if*

$$'M_\beta = M_\beta - \psi(x)_{,\beta}, \tag{4.4}$$

where the generator $\psi(x)$ is an arbitrary function of position of class C^1.

These definitions are analogous to the traditional definitions of conformal and gauge transformations if $I_{\alpha\beta}$ is identified with the coefficient tensor of the fundamental quadratic differential form in the well-based space $\mathfrak{B}$ and if M_β is identified with the coefficient vector of the fundamental linear differential form.‡ As pointed out in Chapter III, however, the field $I_{\alpha\beta}$ has, in general, no preassigned transformation properties. It is only under the assumption of particular classes of Lagrangian functions that $I_{\alpha\beta}$ has the tensor-transformation property.

† J. A. Schouten: *op. cit.*, p. 133.
‡ *Ibid.*

We shall require an additional type of transformation. Let us denote by $\Gamma^{\alpha}{}_{\beta\gamma}(I_{\xi\eta})$ the Christoffel symbol formed from the field $I_{\alpha\beta}$.

DEFINITION 4.3. *A Christoffel symbol* $\Gamma^{\alpha}{}_{\beta\gamma}(I_{\xi\eta})$ *is said to undergo a projective transformation to an image Christoffel symbol* $'\Gamma^{\alpha}{}_{\beta\gamma}(I_{\xi\eta})$ *if and only if*

$$'\Gamma^{\alpha}{}_{\beta\gamma}(I_{\xi\eta}) = \Gamma^{\alpha}{}_{\beta\gamma}(I_{\xi\eta}) - 2\delta^{\alpha}_{(\beta}\,\Theta_{\gamma)}, \tag{4.5}$$

where the generators Θ_{γ} *are general geometric quantities.*†

At this point we note certain relations that will be needed in the following analysis. From (4.3), we have

$$'I_{\alpha\beta} = \frac{1}{\sigma} I_{\alpha\beta}. \tag{4.6}$$

Since $\Gamma^{\alpha}{}_{\beta\gamma}(I_{\xi\eta})$ is defined by

$$\Gamma^{\alpha}{}_{\beta\gamma}(I_{\xi\eta}) = \frac{I^{\alpha\rho}}{2}\,(\partial_{\beta} I_{\gamma\rho} + \partial_{\gamma} I_{\beta\rho} - \partial_{\rho} I_{\beta\gamma}),$$

under a generalized conformal transformation of $I^{\alpha\beta}$ we obtain

$$\Gamma^{\alpha}{}_{\beta\gamma}(I_{\xi\eta}) = \Gamma^{\alpha}{}_{\beta\gamma}('I_{\xi\eta}) + \tfrac{1}{2}\,(\delta^{\alpha}_{\beta}\,\phi_{,\gamma} + \delta^{\alpha}_{\gamma}\,\phi_{,\beta} - \phi_{,\rho}\,'I^{\rho\alpha}\,'I_{\beta\gamma}), \tag{4.7}$$

where

$$\phi = \ln(\sigma). \tag{4.8}$$

Thus, under generalized conformal transformations of $I^{\alpha\beta}$ and projective transformations of $\Gamma^{\alpha}{}_{\beta\gamma}('I_{\xi\eta})$, we have

$$\Gamma^{\alpha}{}_{\beta\gamma}(I_{\xi\eta}) = {}'\Gamma^{\alpha}{}_{\beta\gamma}('I_{\xi\eta}) + \tfrac{1}{2}\,(\delta^{\alpha}_{\beta}\,\phi_{,\gamma} + \delta^{\alpha}_{\gamma}\,\phi_{,\beta} - \phi_{,\rho}\,'I^{\rho\alpha}\,'I_{\beta\gamma}) + \Theta_{\beta}\,\delta^{\alpha}_{\gamma} + \Theta_{\gamma}\,\delta^{\alpha}_{\beta}, \tag{4.9}$$

so that

$$\Gamma^{\rho}{}_{\rho\gamma}(I_{\xi\eta}) = {}'\Gamma^{\rho}{}_{\rho\gamma}('I_{\xi\eta}) + 2\phi_{,\gamma} + 5\Theta_{\gamma}. \tag{4.10}$$

Let us denote by $L^{\alpha}{}_{\beta\gamma}(\phi, \psi, \Theta_{\mu})$ the image of (3.7) under generalized conformal, generalized gauge, and projective transformations of its argu-

† T. Y. Thomas: *op. cit.*, p. 9.

ments, and denote by $'L^{\alpha}{}_{\beta\gamma}$ the connection (3.7) formed from $'\Gamma^{\alpha}{}_{\beta\gamma}('I_{\xi\eta})$, $'I_{\alpha\beta}$, and $'M_{\beta}$. Using (4.3) through (4.10), by a straightforward calculation we get

$$\begin{aligned} L^{\alpha}{}_{\beta\gamma}(\phi, \psi, \Theta_{\mu}) = {}& 'L^{\alpha}{}_{\beta\gamma} \\ & - \tfrac{1}{2}\,\delta^{\alpha}_{\gamma}\,(3\Theta_{\beta} + \phi_{,\beta} + \psi_{,\beta}) \\ & - \tfrac{1}{2}\,\delta^{\alpha}_{\gamma}\,(3\Theta_{\gamma} + \phi_{,\gamma} + \psi_{,\gamma}) \\ & + \tfrac{1}{2}\,'I^{\rho\alpha}\,'I_{\beta\gamma}\,(5\Theta_{\rho} + \phi_{,\rho} + 3\psi_{,\rho}) \end{aligned} \tag{4.11}$$

provided

$$'P^{\alpha\beta}_{\cdot\cdot\gamma} = \sigma P^{\alpha\beta}_{\cdot\cdot\gamma}. \tag{4.12}$$

Thus, under satisfaction of (4.12), the connection (3.7) is invariant under such transformations, that is,

$$L^{\alpha}{}_{\beta\gamma}(\phi, \psi, \Theta_{\mu}) = 'L^{\alpha}{}_{\beta\gamma},$$

if and only if ψ, ϕ, and Θ_{μ} are such that they satisfy the equations

$$3\Theta_{\beta} + \psi_{,\beta} = -\phi_{,\beta} \tag{4.13}$$

and

$$5\Theta_{\beta} + 3\psi_{,\beta} = -\phi_{,\beta}. \tag{4.14}$$

Solving (4.13) and (4.14) for Θ_{β} and $\psi_{,\beta}$ in terms of ϕ gives

$$\Theta_{\beta} = -\tfrac{1}{2}\,\phi_{,\beta} \tag{4.15}$$

and

$$\psi_{,\beta} = \tfrac{1}{2}\,\phi_{,\beta} \tag{4.16}$$

as the required relations between the quantities Θ_{β}, ψ, and ϕ. From these equations, we see that if any one of the quantities ψ or ϕ is zero, then the remaining ones must also be zero, and if $\Theta_{\beta} = 0$ for a particular value of the index, say β_0, then ϕ and ψ must be independent of the coördinate x^{β_0} if $L^{\alpha}{}_{\rho\gamma}$ is to be invariant. We have thus proved the following result.

THEOREM 4.1. *The image* $L^{\alpha}{}_{\beta\gamma}(\phi, \psi, \Theta_{\beta})$ *of* $L^{\alpha}{}_{\beta\gamma}$ *under generalized conformal transformations* (4.3), *generalized gauge transformations* (4.4), *and projective transformations* (4.5), *for* $I^{\alpha\beta}$ *a member of the nonnull or seminull class, is given by*

$$L^{\alpha}{}_{\beta\gamma}(\phi, \psi, \Theta_{\beta}) = {}'L^{\alpha}{}_{\beta\gamma} - \tfrac{1}{2}\,\delta^{\alpha}_{\gamma}\,(3\Theta_{\beta} + \phi_{,\beta} + \psi_{,\beta})$$
$$- \tfrac{1}{2}\,\delta^{\alpha}_{\beta}\,(3\Theta_{\gamma} + \phi_{,\gamma} + \psi_{,\gamma})$$
$$+ \tfrac{1}{2}\,{}'I^{\rho\alpha}\,{}'I_{\beta\gamma}\,(5\Theta_{\rho} + \phi_{,\rho} + 3\psi_{,\rho}), \tag{4.17}$$

provided that $P^{\alpha\beta}{}_{\gamma}$ *is transformed according to the equation*

$${}'P^{\alpha\beta}_{\cdot\cdot\gamma} = \sigma P^{\alpha\beta}_{\cdot\cdot\gamma}, \tag{4.18}$$

where

$$\phi = \ln \sigma(x). \tag{4.19}$$

Thus, under satisfaction of (4.18), $L^{\alpha}{}_{\beta\gamma}$ *is invariant under simultaneous generalized conformal, gauge, and projective transformations with generators related by the equation*

$$\Theta_{\beta} = -\psi_{,\beta} = -\tfrac{1}{2}\,\phi_{,\beta}. \tag{4.20}$$

In order to point out the significance of Theorem 4.1, we examine the case in which the $P^{\alpha\beta}_{\cdot\cdot\gamma}$ field vanishes. Under this assumption, by equation (3.9) we have

$$A_{\alpha\beta} = -2M_{[\beta,\alpha]}. \tag{4.21}$$

Thus, since $A_{\alpha\beta}$ is interpreted as the Maxwellian field tensor, we may identify M_{β} with one-half of the vector potential f_{β} of the electromagnetic field. The generalized gauge transformation (4.4) then becomes the gauge transformation of the second kind encountered in electromagnetic field theory. Theorem 4.1 tells us that to every gauge transformation of the Maxwellian field there is associated a conformal transformation of the $I^{\alpha\beta}$ field and a projective transformation of the Christoffel symbols $\Gamma^{\alpha}{}_{\beta\gamma}$.

An even more important conclusion of Theorem 4.1 is the following.

THEOREM 4.2. *If the Lagrangian function is independent of the quantities* $Q_{(\rho)}$ *and their derivatives, if* $I^{\alpha\beta}$ *is a member of either the nonnull or the seminull class, and if* $P^{\alpha\beta}_{\cdot\cdot\gamma}$ *transforms according to* (4.18), *then the solution manifold of the variant field equations* (3.8) *is determined only to within the set of all generalized conformal, gauge, and projective transformations with generators satisfying* (4.20).

Proof. Since by hypothesis the Lagrangian function is independent of the quantities $Q_{(\rho)}$ and their derivatives, the only quantities to be varied under axiom F_4 are the $L^{\alpha}{}_{\beta\gamma}$. The $L^{\alpha}{}_{\beta\gamma}$ have been shown, however, to be invariant under the simultaneous conformal, gauge, and projective transformations with generators related by (4.20), provided that $P^{\alpha\beta}_{\cdot\cdot\gamma}$ transforms according to (4.18). Hence, given any solution to the variant field equations obtained from a Lagrangian function satisfying the hypotheses, we can construct a family of new solutions by the above system of conformal, gauge, and projective transformations with generators satisfying (4.20). This is due to the fact that the variations induced in the $L^{\alpha}{}_{\beta\gamma}$ by such transformations vanish by the invariance of the $L^{\alpha}{}_{\beta\gamma}$ established in Theorem 4.1.

An interesting result can be established concerning the transformation of nonnull class fields into seminull class fields.

THEOREM 4.3. *Let the hypotheses of Theorem* 4.2 *be satisfied. Then a necessary and sufficient condition for the existence of a simultaneous conformal, gauge, and projective transformation of any nonnull class* $I^{\alpha\beta}$ *field into a seminull class* $I^{\alpha\beta}$ *field for which the field equations are invariant is*

$$M_{[\alpha,\beta]} = 0. \tag{4.22}$$

Proof. It was proved in Theorem 2.2 that an $I^{\alpha\beta}$ field with nonvanishing determinant is a member of the nonnull class if $M_\alpha \neq 0$ and is a member of the seminull class if $M_\alpha = 0$. Suppose $I^{\alpha\beta}$ is a member of the nonnull class; then $'I^{\alpha\beta}$ is a member of the seminull class if and only if $'M_\alpha = 0$. By (4.4), $'M_\alpha$ vanishes if and only if

$$M_\alpha = \psi_{,\alpha}.$$

This equation is completely integrable for ψ if and only if (4.22) is satisfied. If (4.22) is satisfied, then there exists a ψ such that $'M_\alpha = 0$. By Theorem 4.1 and equations (4.20), we can imbed such a gauge transformation of M_α in a simultaneous conformal, gauge, and projective transformation leaving $L^{\alpha}{}_{\beta\gamma}$ invariant and hence, by Theorem 4.3, leaving the field equations invariant.

5. The Case of Scalar Density Lagrangian Functions

Under the assumption that the Lagrangian function is a scalar density, the results of the preceding sections take certain particular forms that will be needed in succeeding chapters. We shall also assume that

$$P^{\alpha\beta}_{\cdot\cdot\gamma} = 0.$$

For $\mathfrak{L}$ a scalar density, (1.5) states that the field $I^{\alpha\beta}$ is a tensor density of contravariant rank 2 and weight 1. Thus the colon derivative, as defined by (1.8), becomes the semicolon (covariant) derivative.

Let $h_{\alpha\beta}$ be a symmetric tensor with nonvanishing determinant h, and let $I^{\alpha\beta}$ be a member of the nonnull or the seminull class; then we may represent $I^{\alpha\beta}$ as

$$I^{\alpha\beta} = a\sqrt{|h|}\, h^{\alpha\beta}, \qquad a = \text{constant}, \tag{5.1}$$

where $h^{\alpha\beta}$ is the unique reciprocal of $h_{\alpha\beta}$, that is,

$$h\, h^{\alpha\beta} = h_{,h_{\alpha\beta}}. \tag{5.2}$$

Equation (5.1) then gives

$$I_{\alpha\beta} = \frac{h_{\alpha\beta}}{a\sqrt{|h|}}. \tag{5.3}$$

Substituting (5.3) into (3.4), we have

$$\Gamma^{\alpha}{}_{\beta\gamma}(I_{\xi\eta}) = \Gamma^{\alpha}{}_{\beta\gamma}(h_{\xi\eta}) - \tfrac{1}{4}\,[(\ln|h|)_{,\gamma}\,\delta^{\alpha}_{\beta} + (\ln|h|)_{,\beta}\,\delta^{\alpha}_{\gamma} - (\ln|h|)_{,\sigma}\,h^{\sigma\alpha}\,h_{\beta\gamma}]. \tag{5.4}$$

Thus in the case

$$P_{\dot{\alpha}\dot{\beta}\gamma} = 0,$$

(3.7) gives

$$L^{\alpha}{}_{\beta\gamma} = \Gamma^{\alpha}{}_{\beta\gamma}(h_{\xi\eta}) - \tfrac{1}{2}\,[M_{\gamma}\,\delta^{\alpha}_{\beta} + M_{\beta}\,\delta^{\alpha}_{\gamma} - 3M_{\sigma}\,h^{\sigma\alpha}\,h_{\beta\gamma}], \tag{5.5}$$

since

$$\Gamma^{\alpha}{}_{\gamma\alpha}(I_{\xi\eta}) = -\tfrac{1}{2}\,(\ln|h|)_{,\gamma}.$$

By use of Theorem 3.1, we can easily show that (1.2) is equivalent to

$$h_{\alpha\beta;\gamma} = M_{\gamma}\,h_{\alpha\beta} - 2h_{\gamma(\alpha}\,M_{\beta)} = -Q_{\gamma\alpha\beta}. \tag{5.6}$$

Thus *field space is a well-based space for scalar-density Lagrangian functions if* $I^{\alpha\beta}$ *is a member of either the nonnull or seminull class and* $P_{\dot{\alpha}\dot{\beta}\gamma} = 0$.

Under a generalized conformal transformation of $I^{\alpha\beta}$, as defined by equation (4.3), by using (5.1) and (5.3) we obtain

$$'h^{\alpha\beta}\sqrt{|'h|} = \sigma\, h^{\alpha\beta}\sqrt{|h|}. \tag{5.7}$$

Thus, taking determinants of both sides of the above equation and substituting the results back into (5.7) gives

$$'h_{\alpha\beta} = \sigma\, h_{\alpha\beta}, \qquad 'h = \sigma^4\, h, \tag{5.8}$$

as the transformation induced in the $h_{\alpha\beta}$ by a generalized conformal transformation of $I^{\alpha\beta}$. Equation (5.8), however, is just the defining equation of a conformal transformation of a general symmetric tensor. Thus generalized conformal transformations reduce to conformal transformations for scalar-density Lagrangian functions.

Let us denote by ∇_μ the covariant derivative formed with respect to the affinity $\Gamma^\alpha{}_{\beta\gamma}(h_{\lambda\eta})$; that is, ∇_μ is the covariant differential operator that annuls the $h_{\alpha\beta}$ field. By use of (II-2.10) and (II-2.11), we obtain the following evaluation of the tensors $R_{\mu\lambda}$ and $A_{\mu\lambda}$

$$R_{\mu\lambda} = J_{\mu\gamma} - \tfrac{3}{2}\, h_{\mu\lambda}\, \nabla_\eta M^\eta_{\cdot} + \tfrac{3}{2}\, M_\mu\, M_\lambda, \tag{5.9}$$

$$A_{\mu\lambda} = 2M_{[\mu,\lambda]}, \tag{5.10}$$

where

$$M^\eta_{\cdot} = M_\rho\, h^{\eta\rho}, \tag{5.11}$$

and where $J_{\mu\lambda}$ is the contraction of the curvature tensor based on the affinity $\Gamma^\alpha{}_{\beta\gamma}(h_{\lambda\eta})$. (See (II-1.6) and (II-1.7) for definition of $J_{\mu\lambda}$.) Set

$$R = R_{\mu\lambda}\, h^{\mu\lambda}, \qquad J = J_{\mu\lambda}\, h^{\mu\lambda}; \tag{5.12}$$

then we have

$$R_{\mu\lambda} - \tfrac{1}{2}\, R\, h_{\mu\lambda} = J_{\mu\lambda} - \tfrac{1}{2}\, J\, h_{\mu\lambda} + \tfrac{3}{2}\, h_{\mu\lambda}\, \nabla_\eta M^\eta_{\cdot}$$

$$+ \tfrac{3}{2}\, M_\mu\, M_\lambda - \tfrac{3}{4}\, M_\eta\, M^\eta_{\cdot}\, h_{\mu\lambda}. \tag{5.13}$$

The results established in this section will be of considerable use in Part Two.

PART TWO

THE VARIANT FIELD THEORY ANALYSIS OF THE CLASSICAL FIELDS

A CHARACTERISTIC feature of the analysis in Part One is that no constitutive assumptions have been made and that the Lagrangian function has not been particularized in any way other than to require it to be admissible. Although a deterministic system of field equations exhibiting Maxwellian structure has been obtained, together with potential representations for the affine connection for nonnull and seminull fields, the theory is of a structural nature only. The purpose of Part Two is to demonstrate the adequacy of the variant field theory by showing that the classical field theories can be obtained by its use. To accomplish this, it is necessary to consider certain particularized Lagrangian functions. This will be done by explicitly stated constitutive assumptions and definitions, so that the relations between the results in this part and those obtained in Part One can easily be perceived. Since the classical field theories have almost invariably been expressed by generally covariant field equations, the particularizations of the Lagrangian function used to obtain them result in generally covariant equations. No particular claim of originality is made for the Lagrangian functions used, nor is a "unified" field theory proposed that results from THE Lagrangian function. Encouraging results are obtained, though it must be admitted that the idea of finding THE ONE LAGRANGIAN FUNCTION, which would supposedly describe all physical happenings from the microcosm to the macrocosm, hints of the grossest conceit.

Chapter V

Empty and Semiempty Field Space

THE FIRST FIELDS we shall examine are the so-called classical vacuum fields. The precise state of emptiness necessary in order that a field may be classed as a vacuum field is often quickly glossed over in the literature, and consequently the results are usually cloudy or vague. In accordance with the axiomatic method of Part One, we lay down precise definitions of what we call empty space and semiempty space. Although we may not get complete acceptance from all readers on these definitions, they at least make the considerations sufficiently precise to admit definite results corresponding with the usually accepted notions of vacuum fields.

1. Scalar Density Lagrangian Functions in an Empty Field Space

Definitions or assumptions stating the presence or absence of those fields that depend on the Lagrangian function for their definition, or stating the arguments of the Lagrangian function, will be referred to as *constitutive* definitions or assumptions. The reason for referring to them as constitutive is that they place restrictions on the nature of the fields defined over field space rather than placing such restrictions directly on the intrinsic geometric structure of field space, and hence are analogous to the constitutive restrictions encountered in the study of continua.

Let us define exactly what we mean by an empty field space.

CONSTITUTIVE ASSUMPTION 1.1. *A field space $\mathfrak{F}$ will be said to be empty if and only if*

$$S^{\alpha} = H^{\alpha\beta} = Z^{(\rho)} = W^{(\rho)\eta} \equiv 0 \tag{1.1}$$

everywhere in $\mathfrak{F}$ *(that is, the Maxwellian current, the Maxwellian current potential, and the* $Z^{(\rho)}$ *and* $W^{(\rho)\eta}$ *fields vanish identically at all points of* $\mathfrak{F}$*).*

THEOREM 1.1. *The most general form that the Lagrangian function can assume in an empty field space is*

$$\mathfrak{L} = \mathfrak{L}(R_{\alpha\beta}, L^{\alpha}{}_{\beta\gamma}, \mathbf{x}). \tag{1.2}$$

Proof. By definition of the fields $H^{\alpha\beta}$, $Z^{(\rho)}$, and $W^{(\rho)\eta}$, we have

$$H^{\alpha\beta} = \mathfrak{L}_{,A_{\alpha\beta}}, \qquad Z^{(\rho)} = \mathfrak{L}_{,Q_{(\rho)}}, \qquad W^{(\rho)\eta} = \mathfrak{L}_{,Q_{(\rho),\eta}}.$$

Thus, by Constitutive Definition 1.1, a field space is empty only if the Lagrangian function is such that

$$0 \equiv \mathfrak{L}_{,A_{\alpha\beta}} = \mathfrak{L}_{,Q_{(\rho)}} = \mathfrak{L}_{,Q_{(\rho),\eta}}.$$

Integrating this system of equations and remembering that by axiom F_4 the arguments of the Lagrangian function are $R_{\alpha\beta}$, $A_{\alpha\beta}$, $L^{\alpha}{}_{\beta\gamma}$, $Q_{(\rho)}$, $Q_{(\rho),\eta}$, and $\mathbf{x}$, we obtain (1.2).

CONSTITUTIVE ASSUMPTION 1.1. *The Lagrangian function is a scalar density such that*

$$I = \det(I^{\alpha\beta}) \neq 0, \tag{1.3}$$

$$\partial_{L^{\alpha}{}_{\beta\gamma}} \mathfrak{L} = 0. \tag{1.4}$$

With this assumption we may state the following result.

THEOREM 1.2. *If* $\mathfrak{F}$ *is an empty field space for which Constitutive Assumption* 1.1 *is satisfied, then the field* $I^{\alpha\beta}$ *is a tensor density and is a member of the seminull class.*

Proof. By (1.4) and the definition of the Maxwellian current S^{α}, we have

$$S^{\alpha} = \tfrac{3}{10} I^{\alpha\eta}{}_{:\eta}.$$

Since S^{α} must vanish in order for the field space to be empty, the above equation yields

$$0 = I^{\alpha\eta}{}_{:\eta}.$$

The defining property (IV-2.1) of a seminull class field, namely

$$I \neq 0, \qquad I^{\alpha\eta}{}_{:\eta} = 0,$$

is thus satisfied. The demonstration that $I^{\alpha\beta}$ is a tensor density under the hypothesis is trivial and is left to the reader.

Note. Since $I^{\alpha\beta}$ has been shown to be a tensor density, covariant differentiation is identical with colon differentiation:

$$I^{\alpha\beta}{}_{:\gamma} \equiv I^{\alpha\beta}{}_{;\gamma}. \tag{1.5}$$

Under the axioms of field space, we derived the system of partial differential equations (III-2.13), the variant field equations, which govern the quantities $I^{\alpha\beta}$, $H^{\alpha\beta}$, $Z^{(\rho)}$, and $W^{(\rho)\eta}$ in such a space. The structure of these field equations is significantly simplified under the constitutive definition and assumption made above.

THEOREM 1.3. *If* $\mathfrak{F}$ *is an empty space for which Constitutive Assumption* 1.1 *is satisfied, then the only variant field equations not identically satisfied are*

$$I^{\alpha\beta}{}_{;\gamma} = 0. \tag{1.6}$$

Proof. The variant field equations, as developed in III-2, are

$$Z^{(\rho)} - W^{(\rho)\eta}{}_{,\eta} = 0, \tag{a}$$

$$H^{\alpha\beta}{}_{;\beta} = S^{\alpha}, \tag{b}$$

$$I^{\alpha\beta}{}_{:\gamma} = \tfrac{2}{5}\,\delta^{(\alpha}_{\gamma}\, I^{\beta)\eta}{}_{:\eta} + P^{\alpha\beta}_{\cdot\cdot\gamma}. \tag{c}$$

Under the definition of empty field space, equations (a) and (b) are identically satisfied. It was shown in the proof of Theorem 1.2 that

$$I^{\alpha\eta}{}_{:\eta} = 0,$$

and by (III-2.12) and (1.4) we have

$$P^{\alpha\beta}_{\cdot\cdot\gamma} = 0.$$

Hence we obtain

$$I^{\alpha\beta}{}_{:\gamma} = 0.$$

Now the colon derivative is the same as the covariant derivative, since $I^{\alpha\beta}$ has been shown to be a tensor density under Constitutive Assumption 1.1, and therefore (1.6) follows.

Note. If we require $\mathfrak{F}$ to be an empty field space with Lagrangian function satisfying (1.3) and (1.4), but not necessarily a scalar density, then Theorem 1.3 still holds provided (1.6) is replaced by

$$I^{\alpha\beta}{}_{:\gamma} = 0.$$

This is easily seen by noting that the proof of Theorem 1.3 continues to hold under the above hypothesis except that the colon derivative can no longer be replaced by the covariant derivative.

Having demonstrated that $I^{\alpha\beta}$ is a member of the seminull class, we see that the potential-form representations of $L^{\alpha}{}_{\beta\gamma}$ obtained in Chapter IV are applicable. We shall not use these results, however, since there is an alternative and direct development that is of interest in its own right and that tends to clarify certain points through its direct approach.

THEOREM 1.4. *If $\mathfrak{F}$ is an empty field space for which Constitutive Assumption 1.1 is satisfied, then $\mathfrak{F}$ is a metric space with metric tensor $h_{\alpha\beta}$ defined by*

$$h^{\alpha\beta} = I^{-1/2}\, I^{\alpha\beta}, \qquad h\, h^{\alpha\beta} = h_{,h_{\alpha\beta}}, \qquad h = \det(h_{\alpha\beta}). \tag{1.7}$$

Hence $L^{\alpha}{}_{\beta\gamma}$ is given by

$$L^{\alpha}{}_{\beta\gamma} = \Gamma^{\alpha}{}_{\beta\gamma}(h_{\eta\pi}), \tag{1.8}$$

where

$$\Gamma^{\alpha}{}_{\beta\gamma}(h_{\eta\pi}) \overset{\text{def}}{=} \tfrac{1}{2}\, h^{\alpha\rho}\, (h_{\rho\beta,\gamma} + h_{\rho\gamma,\beta} - h_{\beta\gamma,\rho}) \tag{1.9}$$

are the Christoffel symbols of the second kind formed from $h_{\alpha\beta}$.

Proof. Since $I^{\alpha\beta}$ is a symmetric tensor density, the determinant of which has been assumed to be nonzero, there exists a unique relative tensor $I_{\alpha\beta}$ of weight -1, defined by

$$I\, I_{\alpha\beta} \overset{\text{def}}{=} I_{,I^{\alpha\beta}}, \tag{1.10}$$

with the property

$$I^{\alpha\beta}\, I_{\beta\gamma} = \delta^{\alpha}_{\gamma}. \tag{1.11}$$

Covariant differentiation of (1.11) leads directly to

$$I^{\alpha\beta}{}_{;\gamma} = -I^{\alpha\eta}\, I^{\beta\mu}\, I_{\eta\mu;\gamma}, \tag{1.12}$$

and hence, since $I \neq 0$, the variant field equations (1.6) are equivalent to

$$I_{\alpha\beta;\gamma} = 0. \tag{1.13}$$

Since the field $I_{\alpha\beta}$ is a relative tensor of weight -1, it can be represented by

$$I_{\alpha\beta} = h_{\alpha\beta}\, h^{-1/2}, \tag{1.14}$$

where $h_{\alpha\beta}$ is a symmetric tensor and

$$h = \det(h_{\alpha\beta}).$$

Taking determinants of both sides of (1.14) gives

$$\det(I_{\alpha\beta}) = [\det(I^{\alpha\beta})]^{-1} = I^{-1} = h^{-1}, \tag{1.15}$$

and hence $h \neq 0$ since $I \neq 0$. Substituting (1.14) into (1.13) gives

$$\frac{h_{\alpha\beta;\gamma}}{h^{1/2}} - \frac{1}{2}\frac{h_{;\gamma}}{h^{3/2}}\, h_{\alpha\beta} = 0 \tag{1.16}$$

as the form of the variant field equations in terms of the $h_{\alpha\beta}$ field. Now, since the chain rule holds for covariant differentiation of determinants of relative tensors, by the definition of $I_{\alpha\beta}$ we have

$$I_{;\gamma} = I_{,I^{\alpha\beta}}\, I^{\alpha\beta}{}_{;\gamma} = I\, I_{\alpha\beta}\, I^{\alpha\beta}{}_{;\gamma}.$$

From Theorem 1.3, we obtain

$$I^{\alpha\beta}{}_{;\gamma} = 0,$$

so that the above calculation of $I_{;\gamma}$ leads to

$$I_{;\gamma} = 0,$$

and hence

$$h_{;\gamma} = 0$$

by (1.15). Thus equation (1.16) yields

$$h_{\alpha\beta;\gamma} = 0 \tag{1.17}$$

as the final form of the variant field equations in empty space under Constitutive Assumption 1.1. Noting that $h_{\alpha\beta}$ is a symmetric tensor with vanishing covariant derivative, and that $h \neq 0$, we see that the postulates M_1, M_2, M_3 of a metric space are satisfied, and hence that the components of affine connection are given by (1.8) and (1.9). It now remains to demonstrate $(1.7)_1$. By equation (1.15), we may eliminate h in (1.14) to obtain

$$h_{\alpha\beta} = I^{1/2}\, I_{\alpha\beta}.$$

Inverting this equation by use of the definitions of $I_{\alpha\beta}$ and the reciprocal field, $h^{\alpha\beta}$, of $h_{\alpha\beta}$, that is, the field satisfying

$$h\, h^{\alpha\beta} \stackrel{\text{def}}{=} h_{,h_{\alpha\beta}},$$

we obtain $(1.7)_1$ as desired.

It is to be noted that we have derived, not assumed, metric structure for empty field space subject to Constitutive Assumption 1.1. The tensor $h_{\alpha\beta}$, thereby proved to be the metric tensor, is uniquely related to the Lagrangian function by $(1.7)_1$ and hence cannot be chosen arbitrarily by a direct assumption of the metric structure of space. Such metric structure must be determined by $(1.7)_1$. To see this, we first note that $I^{\alpha\beta}$ is, by definition, the partial derivative of the Lagrangian function with respect to the field $R_{\alpha\beta}$, and hence will, in general, be a function of $R_{\alpha\beta}$. In a metric space, the field $R_{\alpha\beta}$ is given by

$$R_{\alpha\beta} = J_{\alpha\beta} \stackrel{\text{def}}{=} 2\Gamma^{\eta}{}_{\alpha[\eta,\beta]} + 2\Gamma^{\eta}{}_{\mu[\beta}\, \Gamma_{\eta]}{}^{\mu}{}_{\alpha}, \tag{1.18}$$

and hence is a second-order differential operator on $h_{\alpha\beta}$ since the Christoffel symbols are first-order differential operators (see equation (1.9)). Hence $I^{\alpha\beta}$, and consequently the right-hand side of $(1.7)_1$, is in general a second-order differential operator on $h_{\alpha\beta}$, and thus only those metric tensors $h_{\alpha\beta}$ satisfying the second-order system of partial differential equations $(1.7)_1$ are admissible in an empty field space under Constitutive Assumption 1.1. Summarizing, we have the following result.

THEOREM 1.5. *If $\mathfrak{F}$ is an empty field space for which Constitutive Assumption 1.1 is satisfied, then $\mathfrak{F}$ is a metric space with metric tensor $h_{\alpha\beta}$ satisfying the system of second-order partial differential equations*

$$h^{\alpha\beta} = I^{-1/2}\, I^{\alpha\beta}, \qquad I^{\alpha\beta} = \mathfrak{L}(J_{\mu\nu}, x)_{,J_{\alpha\beta}}. \tag{1.19}$$

2. The Einstein Vacuum Gravitation Equations with Cosmological Constant

Let the ground space be an empty field space with a scalar density Lagrangian function. By Theorem 1.1, the most general form of the Lagrangian function is

$$\mathfrak{L} = \mathfrak{L}(R_{\alpha\beta}, L^{\alpha}{}_{\beta\gamma}, x). \tag{2.1}$$

Under Constitutive Assumption 1.1, we are left with the possible arguments $R_{\alpha\beta}$ and $\mathbf{x}$. The simplest scalar density function that can be formed from these arguments is the square root of the determinant of $R_{\alpha\beta}$. Hence, if we restrict our attention to the simplest form for the Lagrangian function, we are naturally led to select

$$\mathfrak{L} = 2\sqrt{\det(R_{\alpha\beta})}, \tag{2.2}$$

where the factor 2 is so chosen as to simplify the occurrence of numerical factors. To make the analysis nontrivial, we explicitly assume that

$$\det(R_{\alpha\beta}) \neq 0.$$

We now have to determine $I^{\alpha\beta}$ as a function of $R_{\alpha\beta}$. Set

$$\tilde{R} \stackrel{\text{def}}{=} \det(R_{\alpha\beta}), \tag{2.3}$$

so that

$$\mathfrak{L} = 2\tilde{R}^{1/2}. \tag{2.4}$$

Since $\tilde{R}$ has been assumed to be nonzero and $R_{\alpha\beta}$ is a symmetric tensor, there exists a unique reciprocal tensor $R^{\alpha\beta}$, defined by

$$\tilde{R}\, R^{\alpha\beta} \stackrel{\text{def}}{=} \tilde{R}_{,R_{\alpha\beta}}, \tag{2.5}$$

with the property that

$$R^{\alpha\beta} R_{\beta\gamma} = \delta^{\alpha}_{\gamma}. \tag{2.6}$$

Note. $R^{\alpha\beta}$ may be interpreted geometrically as a radius of curvature tensor.

By definition, we have

$$I^{\alpha\beta} = \mathfrak{L}_{,R_{\alpha\beta}}. \tag{2.7}$$

By substitution of (2.4) into (2.7) and use of (2.5) we reduce $I^{\alpha\beta}$ to the explicit form

$$I^{\alpha\beta} = \tilde{R}^{1/2} R^{\alpha\beta}. \tag{2.8}$$

We have assumed that the ground space is an empty field space, and by inspection the Lagrangian function (2.4) satisfies Constitutive Assumption 1.1. It thus remains only to show that $I \neq 0$ to ensure that the results of Section 1 carry over *in toto*. Taking the determinants of both sides of (2.8), we have

$$I = \tilde{R}, \tag{2.9}$$

and hence $I \neq 0$ since $\tilde{R}$ has been assumed to be nonzero. Thus by Theorem 1.5 the space considered is metric, with metric tensor $h_{\alpha\beta}$ satisfying the system of second-order partial differential equations

$$h^{\alpha\beta} = I^{-1/2} I^{\alpha\beta}. \tag{2.10}$$

Substituting (2.8) and (2.9) into (2.10), we obtain

$$h^{\alpha\beta} = R^{\alpha\beta},$$

and hence, upon inversion,

$$h_{\alpha\beta} = R_{\alpha\beta} = J_{\alpha\beta} = 2\Gamma^{\eta}{}_{\alpha[\eta,\beta]} + 2\Gamma^{\eta}{}_{\mu[\beta}\Gamma^{\mu}{}_{\eta]\alpha}, \tag{2.11}$$

since the ground space is metric (see II-1.7). We have thus proved the following result.

THEOREM 2.1. *If $\mathfrak{F}$ is an empty field space with Lagrangian function* (2.4), *then $\mathfrak{F}$ is a metric space with metric tensor $h_{\alpha\beta}$ satisfying the system of second-order partial differential equations*

$$J_{\alpha\beta} = h_{\alpha\beta}. \tag{2.12}$$

By the definition of the Christoffel symbols,

$$\Gamma^{\alpha}{}_{\beta\gamma} = \tfrac{1}{2} h^{\alpha\rho} (h_{\rho\beta,\gamma} + h_{\rho\gamma,\beta} - h_{\beta\gamma,\rho}),$$

the $\Gamma^{\alpha}{}_{\beta\gamma}$ are homogeneous functions of degree 0 in the $h_{\alpha\beta}$ and their derivatives. Thus, by (2.11), the $J_{\alpha\beta}$ are homogeneous functions of degree 0 in the $h_{\alpha\beta}$ and their derivatives. Setting

$$\Lambda\, 'h_{\alpha\beta} = h_{\alpha\beta}, \qquad \Lambda = \text{constant},$$

in (2.12) and dropping the primes, since we have

$$'J_{\alpha\beta} = J_{\alpha\beta}$$

by the above homogeneity argument, we obtain

$$J_{\alpha\beta} = \Lambda\, h_{\alpha\beta}. \tag{2.13}$$

Equations (2.13) are just the Einstein vacuum-gravitation equations with cosmological constant Λ. It might be thought that going from (2.12) to (2.13) by the transformation

$$\Lambda\, 'h_{\alpha\beta} = h_{\alpha\beta}$$

would imply that the metric of physical space must be subjected to such a transformation. This is not so. The metric tensor $h_{\alpha\beta}$ is a derived quantity and has no physical interpretation until it has been tied to "physical" space structure by identification with a physical metric through predictions of the field equations. Now equations (2.13) provide such an identification to the same extent as do the Einstein equations, and this identification results after the transformation in question has been made. It must be noted that we have *derived* the metric structure of space and the Einstein vacuum field equations with cosmological constant directly from the four axioms of field space by means of certain constitutive assumptions and definitions. It is also to be noted that the cosmological constant cannot be zero under the above analysis, for then we would have

$$0 = J_{\alpha\beta} = R_{\alpha\beta},$$

and the assumption $\tilde{R} \neq 0$ would be violated. In this sense we have obtained the cosmological constant in a direct way, and its existence is an essential consequence of the theory for the given Lagrangian function, not something that is left to the taste of the investigator.

THEOREM 2.2. *If $\mathfrak{F}$ is an empty field space with Lagrangian function* (2.4), *then $\mathfrak{F}$ is a metric space such that the Einstein vacuum-gravitation equations with cosmological constant hold.*

It is of interest to note that (2.13) exhibits the following† nonsingular solution:

† G. C. McVittie: *General Relativity and Cosmology*, Chapman and Hall, Ltd., London (1956), p. 73.

$$h_{11} = h_{22} = h_{33} = -\exp(-2x^{(4)}\sqrt{\Lambda/3}), \qquad h_{44} = c^2.$$

Thus the space resulting from the variant field equations in empty field space under (2.4) is approximately a Minkowski space for sufficiently small Λ and for $x^{(4)}$ restricted to a small enough interval.

If we require the variant field equations to be satisfied everywhere except in the vicinity of the origin, we obtain† the solution

$$h_{11} = h_{22} = h_{33} = \frac{-c^2}{h_{44}} = \frac{-1}{1 - 2m/r - \Lambda\, r^2/3},$$

which, of course, represents empty space except in the vicinity of the origin. If we add the ad hoc requirement that test particles describe geodesics and photons describe null-geodesics of this metric space, we obtain the three classical tests of Einstein's general theory.‡ It must be noted that the above theory is for empty field space, in which particles are not described. It is for this reason that their description must be added as ad hoc data. This situation is remedied in Chapters VI through VIII, where the Einstein equations of the gravitational field are obtained, together with equations describing the motion of matter and of test particles.

3. Scalar Density Lagrangian Functions in a Semiempty Field Space

We first define exactly what we mean by a semiempty field space.

CONSTITUTIVE DEFINITION 3.1. *A field space $\mathfrak{F}$ will be said to be a semiempty field space if and only if*

$$S^\alpha = Z^{(\rho)} = W^{(\rho)\eta} \equiv 0 \tag{3.1}$$

everywhere in $\mathfrak{F}$ (that is, the Maxwellian current and the $Z^{(\rho)}$ and $W^{(\rho)\eta}$ fields vanish identically at all points in $\mathfrak{F}$).

With this definition we have the following result.

THEOREM 3.1. *The most general form that the Lagrangian function can assume in a semiempty field space is*

† R. C. Tolman: *Relativity, Thermodynamics, and Cosmology*, Clarendon Press, Oxford (1934), p. 245.

‡ P. G. Bergmann: *Introduction to the Theory of Relativity*, Prentice-Hall, Inc., Englewood Cliffs, N.J. (1942), pp. 211–222.

$$\mathfrak{L} = \mathfrak{L}(R_{\alpha\beta}, A_{\alpha\beta}, L^{\alpha}{}_{\beta\gamma}, x). \tag{3.2}$$

Proof. The proof is identical in structure to that of Theorem 1.1 and is left to the reader.

Constitutive Assumption 3.1. *The Lagrangian function is a scalar density such that*

$$\partial_{L^{\alpha}{}_{\beta\gamma}} \mathfrak{L} = 0. \tag{3.3}$$

With this assumption we may state the following result.

Theorem 3.2. *If $\mathfrak{F}$ is a semiempty field space for which Constitutive Assumption* 3.1 *is satisfied, then $I^{\alpha\beta}$ and $H^{\alpha\beta}$ are tensor densities and*

$$0 = I^{\alpha\eta}{}_{:\eta}, \tag{3.4}$$

so that $I^{\alpha\beta}$ is a member of either the seminull class, the proper null class, or the improper null class.

Proof. That $I^{\alpha\beta}$ and $H^{\alpha\beta}$ are tensor densities under the hypothesis is trivial and is left to the reader. By definition of the Maxwellian current S^{α} and (3.3), we have

$$S^{\alpha} = \tfrac{3}{10} I^{\alpha\eta}{}_{:\eta}.$$

Since S^{α} must vanish in order for the field space to be semiempty, (3.4) must hold. Hence $I^{\alpha\beta}$ is not a member of the nonnull class; since $I^{\alpha\eta}{}_{:\eta}$ must be nonzero for all members of that class. Since our four classes of $I^{\alpha\beta}$ fields are exhaustive, the result follows.

Note 1. If $I^{\alpha\beta} = 0$, then $I^{\alpha\beta}$ is a member of the proper null class, and if $I \neq 0$, then, by (3.4), $I^{\alpha\beta}$ is a member of the seminull class. If, on the other hand, $I = 0$ and $I^{\alpha\beta} \neq 0$ for some (α, β), then $I^{\alpha\beta}$ is a member of the improper null class.

Note 2. Since $I^{\alpha\beta}$ has been shown to be a tensor density, the colon derivative is the same as the covariant derivative.

The structure of the variant field equations is somewhat simplified under the constitutive assumption and definition made above, as might be expected.

Theorem 3.3. *If $\mathfrak{F}$ is a semiempty field space for which Constitutive Assumption* 3.1 *is satisfied, then the only variant field equations not identically satisfied are*

$$I^{\alpha\beta}{}_{;\gamma} = 0, \tag{3.5}$$

$$H^{\alpha\beta}{}_{,\beta} = 0. \tag{3.6}$$

Proof. The variant field equations, as developed in III-2, are

$$Z^{(\rho)} - W^{(\rho)\eta}{}_{,\eta} = 0, \tag{a}$$

$$H^{\alpha\beta}{}_{,\beta} = S^{\alpha}, \tag{b}$$

$$I^{\alpha\beta}{}_{:\gamma} = \tfrac{2}{5}\,\delta^{(\alpha}_{\gamma}\, I^{\beta)\eta}{}_{:\eta} + P^{\alpha\beta}_{\cdot\,\cdot\,\gamma}. \tag{c}$$

Under the definition of semiempty field space, equation (a) is identically satisfied and equation (b) reduces to (3.6). It was shown in the proof of Theorem 3.2 that

$$I^{\alpha\eta}{}_{:\eta} = 0,$$

so that (c) becomes

$$I^{\alpha\beta}{}_{:\gamma} = 0,$$

as a result of the fact that, by (III-2.12) and (3.3),

$$P^{\alpha\beta}_{\cdot\,\cdot\,\gamma} = 0.$$

Equation (3.5) thus follows since the colon derivative is the same as the covariant derivative, as a result of the already established fact that $I^{\alpha\beta}$ is a tensor density under Constitutive Assumption 3.1.

Since it has been demonstrated that $I^{\alpha\beta}$ is not a member of the nonnull class, the potential-form representations of $L^{\alpha}{}_{\beta\gamma}$, obtained in Chapter IV, are not generally applicable. The solution for the affine connection is straightforward, however, though lengthy. We begin with the following result.

THEOREM 3.4. *If $\mathfrak{F}$ is a semiempty field space for which Constitutive Assumption 3.1 is satisfied, and if the Lagrangian function is such that $I \neq 0$, then $\mathfrak{F}$ is a metric space, with metric tensor $h_{\alpha\beta}$ defined by*

$$h^{\alpha\beta} = I^{-1/2}\, I^{\alpha\beta}, \qquad h\, h^{\alpha\beta} = h_{,h_{\alpha\beta}}, \qquad h = \det(h_{\alpha\beta}) \neq 0, \tag{3.7}$$

and hence $L^{\alpha}{}_{\beta\gamma}$ is given by

$$L^{\alpha}{}_{\beta\gamma} = \Gamma^{\alpha}{}_{\beta\gamma}(h_{\eta\mu}). \tag{3.8}$$

Proof. By hypothesis, we have $I \neq 0$. This condition, together with Constitutive Assumption 3.1, implies that Constitutive Assumption 1.1 is also satisfied. We have shown in Theorem 3.3 that the equation

$$I^{\alpha\beta}{}_{;\gamma} = 0$$

must hold in semiempty field space under Constitutive Assumption 3.1. If we examine the proof of Theorem 1.4, however, we see that only the data established to this point in the current proof are used there. Thus Theorem 1.4 holds, establishing the desired result.

Theorem 3.4 implies a result we shall require later in contexts other than semiempty field space. For this reason we give the following equivalent statement.

THEOREM 3.4′. *Let $\mathfrak{F}$ be a field space such that $I^{\alpha\beta}$ is a tensor density and such that*

$$P^{\alpha\beta}_{\cdot\cdot\gamma} = 0.$$

If

$$I \neq 0 \quad \textit{and} \quad I^{\alpha\beta}{}_{;\beta} = 0,$$

then $\mathfrak{F}$ is a metric space, with metric tensor $h_{\alpha\beta}$ defined by (3.7), *and hence $L^{\alpha}{}_{\beta\gamma}$ is given by*

$$L^{\alpha}{}_{\beta\gamma} = \Gamma^{\alpha}{}_{\beta\gamma}(h_{\eta\mu}).$$

Proof. Substituting

$$I^{\alpha\beta}{}_{:\beta} = 0$$

into the field equation

$$I^{\alpha\beta}{}_{:\gamma} = \tfrac{2}{5}\,\delta^{(\alpha}_{\gamma}\, I^{\beta)\eta}{}_{:\eta} + P^{\alpha\beta}_{\cdot\cdot\gamma}$$

and noting that

$$P^{\alpha\beta}_{\cdot\cdot\gamma} = 0$$

by hypothesis, we have

$$I^{\alpha\beta}{}_{:\gamma} = 0.$$

This result, together with the assumption $I \neq 0$, is equivalent to the necessary hypotheses of Theorem 3.4, and hence the theorem follows.

Theorem 3.4 completes the case $I \neq 0$, but is somewhat uninteresting in that no new results are obtained. The main purpose of Theorem 3.4 is to allow us to establish the following result.

THEOREM 3.5. *Let $\mathfrak{F}$ be a semiempty space in which Constitutive Assumption* 3.1 *is satisfied. If $I \neq 0$ then the Maxwellian field tensor $A_{\alpha\beta}$ vanishes everywhere in $\mathfrak{F}$; conversely, $A_{\alpha\beta} \neq 0$ for some (α, β) only if $I = 0$.*

Proof. By Theorem 3.4, the ground space is a metric space if $I \neq 0$. We have shown in Theorem II-2.2 that a necessary condition for the reduction of an affinely connected space, and hence a field space, to a metric space is $A_{\alpha\beta} = 0$. Thus, since $I \neq 0$ implies the reduction of semiempty field space to a metric space, we must have $A_{\alpha\beta} = 0$ everywhere in $\mathfrak{F}$ if $I \neq 0$. By definition, $A_{\alpha\beta}$ satisfies the equation

$$A_{\alpha\beta} = 2L^{\eta}{}_{\eta[\beta,\alpha]}.$$

Hence, in order to have $A_{\alpha\beta} \neq 0$ for some (α, β), we must have

$$L^{\eta}{}_{\eta[\beta,\alpha]} \neq 0$$

for some (α, β). In a metric space, $L^{\alpha}{}_{\beta\gamma}$ is given by

$$L^{\alpha}{}_{\beta\gamma} = \Gamma^{\alpha}{}_{\beta\gamma}(h_{\eta\mu}),$$

so that

$$L^{\eta}{}_{\eta\beta} = \Gamma^{\eta}{}_{\eta\beta}(h_{\eta\mu}) = \tfrac{1}{2}\,(\ln |h|)_{,\beta}$$

and hence

$$L^{\eta}{}_{\eta[\beta,\alpha]} = 0$$

in such a space. Thus semiempty field space cannot be a metric space if $A_{\alpha\beta}$ is to be nonzero for some (α, β), and hence I must vanish.

4. Continuation—Proper Semiempty Field Space

Theorems 3.4 and 3.5 show that a semiempty field space effectively reduces to an empty field space if we do not require the Maxwellian field tensor $A_{\alpha\beta}$ to be nonzero for some (α, β).

CONSTITUTIVE DEFINITION 4.1. *A semiempty field space will be said to be proper if and only if $I = 0$, which by Theorem 3.5 is equivalent to requiring $A_{\alpha\beta} \not\equiv 0$ for some (α, β).*

It is of interest to note that this is the first constitutive definition or assumption which has led directly to a requirement on the structure of the differential geometry of the ground space.

THEOREM 4.1. *A necessary and sufficient condition for a semiempty field space to be proper is that there exists a nonzero vector function f_β, which is not a gradient vector, such that*

$$L^{\eta}{}_{\eta\beta} = -\tfrac{1}{2} f_\beta. \tag{4.1}$$

Proof. In order for a semiempty field space to be proper, we must, by definition, have

$$A_{\alpha\beta} \not\equiv 0 \tag{4.2}$$

for some (α, β). By (II-3.4) and the fact that the ground space is a field space, we have

$$A_{[\alpha\beta,\gamma]} = 0,$$

and hence there exists a vector f_α such that

$$f_{[\alpha,\beta]} = A_{\alpha\beta}. \tag{4.3}$$

Hence, we have $A_{\alpha\beta} \not\equiv 0$ for some (α, β) if and only if

$$f_{[\alpha,\beta]} \not\equiv 0$$

for some (α, β). This condition can be satisfied if and only if f_α is a nonzero vector but not a gradient vector. Expanding the right-hand side of (4.3) by means of the definition (II-2.5) of $A_{\alpha\beta}$ gives

$$f_{[\alpha,\beta]} = 2L^{\eta}{}_{\eta[\beta,\alpha]}. \tag{4.4}$$

Integrating (4.4) leads directly to (4.1).

The problem of establishing the form of the affine connection for proper semiempty field spaces can now be expressed as follows: We have to solve the system of field equations $I = 0$,

$$I^{\alpha\beta}{}_{;\gamma} = 0, \tag{4.5}$$

and

$$H^{\alpha\beta}{}_{,\beta} = 0 \tag{4.6}$$

for $L^{\alpha}{}_{\beta\gamma}$, subject to the constraint

$$L^{\eta}{}_{\eta\beta} = -\tfrac{1}{2} f_{\beta}, \tag{4.7}$$

where f_{β} is a nonzero vector but not a gradient vector (see Theorems 3.3 and 4.1). There are basically two ways in which (4.7) can be imbedded in a general representation for the components of affine connection:

$$\text{(a)} \qquad L^{\alpha}{}_{\beta\gamma} = V^{\alpha}{}_{\beta\gamma} - \frac{a}{5}\, \delta^{\alpha}_{(\beta} f_{\gamma)}, \tag{4.8}$$

where $V^{\alpha}{}_{\beta\gamma}$ is an arbitrary symmetric affinity constrained by the conditions

$$V^{\alpha}{}_{\alpha\gamma} = \frac{b}{2} f_{\gamma}, \qquad a = 1 + b, \tag{4.9}$$

$$\text{(b)} \qquad L^{\alpha}{}_{\beta\gamma} = \Gamma^{\alpha}{}_{\beta\gamma}(s_{\eta\mu}) + 2p\, \delta^{\alpha}_{(\beta} f_{\gamma)} + q\, f_{\mu}\, s^{\mu\alpha}\, s_{\beta\gamma}, \tag{4.10}$$

where the $\Gamma^{\alpha}{}_{\beta\gamma}(s_{\eta\mu})$ are Christoffel symbols based on a symmetric tensor $s_{\alpha\beta}$ with nonvanishing determinant, and where

$$5p + q = -\tfrac{1}{2}. \tag{4.11}$$

We have seen that in a proper semiempty space the $I^{\alpha\beta}$ field either is a member of the proper null class, in which case $I^{\alpha\beta} = 0$, or is a member of the improper null class, in which case $I^{\alpha\beta} \neq 0$ for some (α, β) but $I = 0$. The alternative forms (4.8) and (4.10) are just those that are required for these two classes of $I^{\alpha\beta}$ fields.

If $I^{\alpha\beta} \neq 0$ for some (α, β) with $I = 0$, equations (4.5) and (4.6) are 44 equations of which only 40 are independent. Equation (4.5) no longer implies a metric space, since $I = 0$. If the affinity (4.8) is used, the 40 independent field equations together with the 4 conditions $(4.9)_1$ yield 44 equations for the determination of the 44 functions $V^{\alpha}{}_{\beta\gamma}$ and f_{α}. It is evident that, for $I^{\alpha\beta} \neq 0$, the use of (4.10) would yield an overdetermined system since there would then be only 14 functions $s_{\alpha\beta}$ and f_{α} to be determined by the 40 field equations. We thus have the following result.

THEOREM 4.2. *If $\mathfrak{F}$ is a proper semiempty field space for which Constitutive Assumption* 3.1 *is satisfied and $I^{\alpha\beta}$ is a member of the improper null class, then* *the form of the affine connection is given by* (4.8), *where the 44 functions* $V^\alpha{}_{\beta\gamma}$ *and* f_α *are any solutions to* (4.5), (4.6), *and* (4.9).

The case $I^{\alpha\beta} = 0$ now remains to be solved. We first note that $I^{\alpha\beta} = 0$ is an exact formal solution to (4.5). If this formal solution is algebraically consistent, then (4.5) and (4.6) reduce to the 14 equations

$$H^{\alpha\beta}{}_{,\beta} = 0, \tag{4.12}$$

$$I^{\alpha\beta} = 0. \tag{4.13}$$

Using the affinity (4.10), we obtain a deterministic system since (4.12) and (4.13) are 14 equations for the determination of the 14 unknown functions $s_{\alpha\beta}$ and f_γ. On the other hand, if (4.8) were used, we would have only 14 field equations (4.12) and (4.13), plus the 4 equations (4.9), for the determination of the 44 unknown functions $V^\alpha{}_{\beta\gamma}$ and f_α. The field equations would then determine the affinity only to within 30 arbitrary functions of position. We thus have established the following result.

THEOREM 4.3. *If* $\mathfrak{F}$ *is a proper semiempty field space for which Constitutive Assumption* 3.1 *is satisfied and* $I^{\alpha\beta}$ *is a member of the proper null class, then the form of the affine connection is given by* (4.10), *where the 14 functions* $s_{\alpha\beta}$ *and* f_α *are any solutions to* (4.12) *and* (4.13).

It is to be noted, from Theorem 4.3, that the affine connection is not unique since we have free choice of either the constant p or the constant q, the other one being determined by (4.11). We may use this freedom in the choice of p or q to obtain a form of the affine connection that will be required in the next section. Set

$$p = -q = -\tfrac{1}{8},$$

so that (4.11) is identically satisfied. For this choice, (4.10) becomes

$$L^\alpha{}_{\beta\gamma} = \Gamma^\alpha{}_{\beta\gamma}(s_{\mu\eta}) - \tfrac{1}{8}\,(f_\beta\,\delta^\alpha_\gamma + f_\gamma\,\delta^\alpha_\beta - f_\mu\,s^{\mu\alpha}\,s_{\beta\gamma}). \tag{4.14}$$

From (4.14) and (I-5.4) it is evident that the ground space is well based, with base field $s_{\alpha\beta}$, and that

$$Q_{\gamma\alpha\beta} = -\tfrac{1}{4}f_\gamma\,s_{\alpha\beta}. \tag{4.15}$$

From (I-5.3) we thus obtain an evaluation of the covariant derivative of $s_{\alpha\beta}$, namely

$$s_{\alpha\beta;\gamma} = \tfrac{1}{4} f_\gamma \, s_{\alpha\beta}, \tag{4.16}$$

and hence, by (II-2.11), the field tensor $R_{\alpha\beta}$ can be expressed in terms of the f_α, $s_{\alpha\beta}$, and their derivatives. The affine connection determined by (4.16), namely (4.14), is that of Weyl's geometry,† where $s_{\alpha\beta}\, dx^\alpha\, dx^\beta$ is the fundamental quadratic form and $-\frac{1}{4} f_\alpha\, dx^\alpha$ is the fundamental linear form. If $s_{\alpha\beta}$ undergoes a conformal transformation,

$$'s_{\alpha\beta} = \lambda(x)\, s_{\alpha\beta}, \tag{4.17}$$

then we have

$$'s_{\alpha\beta;\gamma} = (\tfrac{1}{4} f_\gamma \lambda + \lambda_{,\gamma})\, s_{\alpha\beta} = (\tfrac{1}{4} f_\gamma + (\ln \lambda)_{,\gamma})\, 's_{\alpha\beta}.$$

Hence, if we use $'s_{\alpha\beta}$ in computing $\Gamma^\alpha{}_{\beta\gamma}$ instead of $s_{\alpha\beta}$, and if at the same time we transform f_α into

$$'f_\alpha = f_\alpha + 4(\ln \lambda)_{,\alpha}, \tag{4.18}$$

then we get the same affinity (4.14). This implies that if (4.14) is used, then the tensor $s_{\alpha\beta}$ is fixed only to within an arbitrary gauge factor $\lambda(x)$, and that f_α is determined only to within the above transformation. This arbitrariness in the choice of gauge is just what should be expected, however, since the Maxwellian field $A_{\alpha\beta}$ must be gauge invariant if it is to correspond to the field tensor of classical electromagnetic theory. Summarizing, we have the following result.

THEOREM 4.4. *If $\mathfrak{F}$ is a proper semiempty field space for which Constitutive Assumption* 3.1 *is satisfied, then one possible (but not unique) representation of the affine connection is given by* (4.14), *where f_γ and $s_{\alpha\beta}$ are any solutions to the potential equations*

$$H^{\alpha\eta}{}_{,\eta} = 0, \qquad I^{\alpha\beta} = 0, \tag{4.19}$$

under which $s_{\alpha\beta}$ is determined only to within an arbitrary gauge factor $\lambda(x)$ and f_γ is determined to within the gauge transformation

$$'f_\alpha = f_\alpha + 4(\ln \lambda)_{,\alpha}. \tag{4.20}$$

Consequently, such a proper semiempty field space exhibits Weyl structure.

† J. A. Schouten: *op. cit.*, p. 133.

5. Vacuum Electrodynamics†

Consider the scalar-density Lagrangian function

$$\mathfrak{L} = \tfrac{1}{2}\sqrt{-h}\,A_{\alpha\beta}\,A_{\gamma\lambda}\,h^{\alpha\gamma}\,h^{\beta\lambda} \tag{5.1}$$

in a general field space, where $h_{\alpha\beta}$ is a symmetric tensor of signature -2,

$$h = \det(h_{\alpha\beta}) < 0, \tag{5.2}$$

and

$$h^{\alpha\beta}\,h = h_{,h_{\alpha\beta}}. \tag{5.3}$$

From the definitions of $H^{\alpha\beta}$, $I^{\alpha\beta}$, $W^{(\rho)}$, $Z^{(\rho)\eta}$, and S^{α}, we have

$$H^{\alpha\beta} = \sqrt{-h}\,A_{\gamma\lambda}\,h^{\alpha\gamma}\,h^{\beta\lambda} \tag{5.4}$$

and

$$S^{\alpha} = I^{\alpha\beta} = Z^{(\rho)} = W^{(\rho)\eta} = 0. \tag{5.5}$$

Under (5.5), our field space becomes a proper semiempty one and $I^{\alpha\beta}$ becomes a member of the proper null class. Further, the function (5.1) obviously satisfies Constitutive Assumption 3.1. Theorem 4.4 is thus applicable, and hence we have

$$L^{\alpha}{}_{\beta\gamma} = \Gamma^{\alpha}{}_{\beta\gamma}(s_{\eta\mu}) - \tfrac{1}{8}\,(f_{\gamma}\,\delta^{\alpha}_{\beta} + f_{\beta}\,\delta^{\alpha}_{\gamma} - f_{\mu}\,s^{\mu\alpha}\,s_{\beta\gamma}), \tag{5.6}$$

where the functions f_{α} and $s_{\alpha\beta}$ are to be determined by the equations

$$H^{\alpha\beta}{}_{,\beta} = 0 \tag{5.7}$$

and

$$I^{\alpha\beta} = 0. \tag{5.8}$$

Under (5.1), however, (5.8) is identically satisfied, and hence there is no constraint on the choice of the functions $s_{\alpha\beta}$. Hence we may set

$$h_{\alpha\beta} = s_{\alpha\beta}$$

with no loss of generality. Since the tensor $h_{\alpha\beta}$ can be chosen at will,

† For a concise treatment in terms of fields in the space-time of special relativity, see A. Mercier: *Analytical and Canonical Formalism in Physics*, North-Holland Publishing Co., Amsterdam (1959), pp. 87–96. For the general case, see A. Peres: "Nonlinear Electrodynamics in General Relativity," *Phys. Rev.*, **122** (1961), pp. 273–275.

subject to the constraint (5.2), and $s_{\alpha\beta}$ has been shown to be the coefficient tensor of the fundamental quadratic form, there is thus no restriction on the metric tensor and hence no restriction other than (5.2) on the choice of coördinate systems for the ground space.

The sole remaining field equations are then the 4 equations

$$H^{\alpha\beta}{}_{,\beta} = 0 \tag{5.9}$$

for the determination of the 4 functions f_α. These field equations, together with (5.4) and

$$A_{\alpha\beta} = f_{[\alpha,\beta]},$$

are just the field equations of the classic vacuum electromagnetic field with vector potential f_α and field tensor $A_{\alpha\beta}$. Moreover, the Lagrangian function (5.1) together with the field tensor $A_{\alpha\beta}$ and the current potentials $H^{\alpha\beta}$ are invariant under all gauge transformations

$$'f_\alpha = f_\alpha + 4(\ln \lambda)_{,\alpha},$$

the Lorentz gauge condition can be used to fix the gauge by solving

$$('f_\beta\, 'h^{\beta\alpha})_{;\alpha} = 0,$$

and $\mathcal{L}_{,h_{\alpha\beta}}$ yields the "field momentum energy tensor"

$$A_{\lambda\eta}\, A_{\rho\pi}\, h^{\lambda\rho}\, (h^{\eta\pi}\, h^{\alpha\beta}/4 - h^{\eta\alpha}\, h^{\pi\beta}).$$

THEOREM 5.1. *The solution manifold of the variant field equations with Lagrangian function* (5.1) *is isomorphic with the solution manifold of classic vacuum electrodynamics under the choice*

$$p = -q = -\tfrac{1}{8}$$

in (4.10). *In addition, the Maxwellian field tensor* $A_{\alpha\beta}$, *the Maxwellian current potentials* $H^{\alpha\beta}$, *and the Maxwellian vector potential* f_α *are, respectively, the field tensor, the current potentials, and the vector potential of classical vacuum electrodynamics.*

By this theorem we have obtained validation of our considerations of Maxwellian fields and the identification of the curvature tensor $A_{\alpha\beta}$ with the Maxwellian field tensor, at least with respect to vacuum electrodynamics.

Another interesting result is seen as follows. The complete system of equations obtained above from considerations in a proper semiempty field space are

$$\mathfrak{L} = \tfrac{1}{2}\sqrt{-h}\, A_{\alpha\beta}\, A_{\delta\gamma}\, h^{\alpha\delta}\, h^{\beta\gamma},$$

$$H^{\alpha\beta} = \sqrt{-h}\, A_{\mu\nu}\, h^{\mu\alpha}\, h^{\beta\gamma},$$

$$H^{\alpha\beta}{}_{,\beta} = 0, \tag{5.10}$$

$$A_{\alpha\beta} = f_{[\alpha,\beta]},$$

in which the last equation is equivalent to

$$A_{[\alpha\beta,\gamma]} = 0.$$

The affine connection in such a proper semiempty field space is

$$L^{\alpha}{}_{\beta\gamma} = \Gamma^{\alpha}{}_{\beta\gamma}(h_{\eta\mu}) - \tfrac{1}{8}\,(f_{\gamma}\,\delta^{\alpha}_{\beta} + f_{\beta}\,\delta^{\alpha}_{\gamma} - f_{\mu}\,h^{\mu\alpha}\,h_{\beta\gamma}), \tag{5.12}$$

so that the *abstract field space* admits Weyl structure. Denote by ∇_β the covariant derivative formed from the affinity $\Gamma^{\alpha}{}_{\beta\gamma}(h_{\mu\eta})$. Since $H^{\alpha\beta}$ is a skew-symmetric field, and is a tensor density by $(5.10)_2$, we have

$$\nabla_\beta H^{\alpha\beta} \stackrel{\text{def}}{=} H^{\alpha\beta}{}_{,\beta} + \Gamma^{\alpha}{}_{\mu\beta}\, H^{\mu\beta} + \Gamma^{\beta}{}_{\mu\beta}\, H^{\alpha\mu} - \Gamma^{\mu}{}_{\mu\beta}\, H^{\alpha\beta} = H^{\alpha\beta}{}_{,\beta},$$

where the last two terms on the right cancel because of the symmetry of $\Gamma^{\alpha}{}_{\beta\gamma}$, and

$$\Gamma^{\alpha}{}_{\mu\beta}\, H^{\mu\beta} = 0$$

because of the symmetry $\Gamma^{\alpha}{}_{\beta\gamma}$ and the skew-symmetry of $H^{\alpha\beta}$. Similarly, it can easily be shown that

$$f_{[\alpha,\beta]} = \nabla_{[\beta} f_{\alpha]}$$

and

$$A_{[\alpha\beta,\gamma]} = \nabla_{[\gamma} A_{\alpha\beta]}.$$

The field equations (5.11) thus assume the equivalent form

$$\nabla_\beta H^{\alpha\beta} = 0, \tag{5.13}$$

$$A_{\alpha\beta} = \nabla_{[\beta} f_{\alpha]}.$$

Equations (5.10) and (5.13) now contain only operations that are well defined in a metric space with metric tensor $h_{\alpha\beta}$, since by definition we have

$$\nabla_\gamma h_{\alpha\beta} = 0.$$

Thus, if we consider a *subsidiary* metric space with metric tensor $h_{\alpha\beta}$, equations (5.10) and (5.13) constitute vacuum electrodynamics in which the gauge invariance of $A_{\alpha\beta}$, $H^{\alpha\beta}$, and $\mathfrak{L}$ still follows and *we do not have to require the field* $h_{\alpha\beta}$ *to undergo a conformal transformation in the presence of a gauge transformation in the* f_β *since the affine connection of a metric space is just* $\Gamma^\alpha{}_{\beta\gamma}(h_{\mu\eta})$, not (5.12). Thus, in this case, *field space may be viewed as an abstract space used for purposes of deriving the governing field equations from an axiomatic basis; once developed, the equations may be taken over into a subsidiary metric space and identified with the field equations of vacuum electrodynamics in a metric space of the Riemannian variety, which is not subjected to conformal transformations of its metric tensor and which admits gauge invariance of the* $A_{\alpha\beta}$, *the* $H^{\alpha\beta}$, *and the Lagrangian function.*

The major criticism of Weyl's noble attempt to put gravitation and electromagnetism on an equal footing was that the space he required for the description of physical phenomena was subject to conformal transformations of its metric tensor, in conflict with the regularity of spectral lines throughout the observed universe. There was no direct way out of this dilemma for Weyl, since he had started with the assumption of metric structure and the identification of the metric with the physical metric; Weyl later stated, however, that his considerations must actually hold in an abstract space that is not the actual physical 4-space of observable events. The derivation given above does not suffer from the same difficulty, since we have not based our field equations on any assumption of the metric structure of space. The $h_{\alpha\beta}$ field is arbitrary in the development, and hence the field equations, once developed in the abstract field space, can be interpreted in a physical metric space by identifying $h_{\alpha\beta}$ with the metric tensor, thereby obtaining classical vacuum electrodynamics with the required gauge invariance but without the problematic conformal changes in the metric tensor.

The results of this section, combined with the results of Section 2, show that for one choice of the Lagrangian function the variant field equations describe the classic vacuum electromagnetic field, and for another choice they describe the classic vacuum gravitational field with cosmological constant. The Lagrangian function thus plays the role of a constitutive function for the gravitational field as well as for the electromagnetic field. This result is analogous to the situation in which the continuum equilibrium equations may describe either a solid or a fluid relative to the choice of constitutive function. We thus see that the *classical vacuum gravitational and vacuum electromagnetic fields are both described by the variant field equations and correspond to a simple separation of effects through appropriate choices of the constitutive (Lagrangian) function for the particular field under investigation. In addition, both fields can be interpreted in a physical metric space of the Riemannian variety so that physical interpretations are well defined.*

Some of the intrinsic differences between vacuum electromagnetic and vacuum gravitational fields can now be explained. We have seen that the electromagnetic field tensor $A_{\alpha\beta}$ is intrinsically associated with the structure of field space in that it is a contraction of the affine curvature tensor. The potential tensor of the gravitational field, however, is not so simply related to curvature structure. In fact, we must compute first and second derivatives of the gravitational potential tensor and combine them in an essentially nonlinear fashion in order to obtain curvature expressions. Viewed the other way, curvature expressions yield second-order, nonlinear differential equations for the determination of the gravitational potential tensor, whereas the electromagnetic field tensor is given directly in terms of curvature expressions. In addition, it has been shown that equations of Maxwellian form are always present for the class of all field theories spanned by the axioms of field space, but those for the gravitational field must be derived under particularization of the Lagrangian function. Thus electromagnetic-like structures are natural, both mathematically and physically, whereas gravitational phenomena, although natural physically, result from the mathematics only under very special assumptions. In this sense, electromagnetic phenomena may be viewed as the physical realization of pure geometry.

It is of interest to note that if we take

$$\mathfrak{L} = 2\sigma\,(-e(\lambda))^{1/2}$$

as the Lagrangian function instead of (5.1), where σ and λ are constants and

$$e(\lambda) = \lambda^2 \sqrt{-h}\,(\lambda^2 + 2A), \qquad 4A = A_{\alpha\beta}\,A_{\gamma\delta}\,h^{\alpha\gamma}\,h^{\beta\delta},$$

we then have

$$H^{\alpha\beta} = -\sigma\,A_{\gamma\delta}\,h^{\alpha\gamma}\,h^{\beta\delta}\sqrt{-h}\,(1 + 2A/\lambda^2)^{-1/2}.$$

As λ approaches infinity, we recover the previous case. For finite λ, we have the Born–Infeld nonlinear electrodynamics with its attendant results.† One may also consider more general Lagrangian functions with asymptotic forms yielding the classical Maxwellian field. Such avenues of approach have been found to be barren in the past, and considering the results to be established in Chapters VII and VIII, it is believed that they are probably unnecessary for adequate descriptions of nature. The significant point to note here is that if one desires to consider such non-linear electromagnetic models, there is no change in the structure of the problem under the variant field theory—just an increase in complexity and difficulty.

† M. Born and L. Infeld: "Foundations of the New Field Theory," *Proc. Roy. Soc. London*, **A144** (1934), pp. 425–451, with the exception of the "derivation of the equations of motion," which is not correct.

Chapter VI

Macroscopic Matter and Noninteracting Gravitational and Electromagnetic Fields

THE PRECEDING CHAPTER has shown that field space yields an adequate description of the vacuum gravitational field and the vacuum electromagnetic field. The next logical step is to inquire whether the same conclusions can be drawn when the space contains macroscopic matter. The present chapter is devoted to answering this question for noninteracting gravitational and electromagnetic fields, and to developing the ground work for the cases in which interaction is allowed.

1. The Basic Field Equations

The structural theory of the variant field arises from the four axioms F_1, F_2, F_3, and F_4 of field space. The first three of these axioms require a 4-dimensional Hausdorff space into which a structure is introduced by requiring that there be defined at every point an affine connection $L^{\alpha}{}_{\beta\gamma}$ with vanishing associated torsion tensor. The curvature tensor $K^{\pi}{}_{\lambda\rho\mu}$ associated with $L^{\alpha}{}_{\beta\gamma}$ is thus well defined,

$$K^{\pi}{}_{\lambda\rho\mu} \stackrel{\text{def}}{=} 2L^{\pi}{}_{\lambda[\mu,\rho]} + 2L^{\pi}{}_{\sigma[\rho}\, L^{\sigma}{}_{\mu]\lambda}, \tag{1.1}$$

and from it the fundamental field tensors

$$A_{\alpha\beta} = -2C_{[\alpha\beta]} = 2L^{\eta}{}_{\eta[\beta,\alpha]} \tag{1.2}$$

and

$$R_{\alpha\beta} = C_{(\alpha\beta)} = L^{\eta}{}_{\eta(\alpha,\beta)} - L^{\eta}{}_{\alpha\beta\,\eta} + 2L^{\mu}{}_{\rho[\alpha}\, L^{\rho}{}_{\mu]\beta} \tag{1.3}$$

are obtained ($C_{\alpha\beta} \stackrel{\text{def}}{=} K^{\eta}{}_{\alpha\beta\eta}$). The fourth axiom of field space requires the existence of a Lagrangian function $\mathfrak{L}(R_{\alpha\beta}, A_{\alpha\beta}, L^{\alpha}{}_{\beta\gamma}, Q_{(\rho)}, Q_{(\rho),\eta}, \mathbf{x})$ of class C^2 in its arguments, and also requires that the functional

$$\mathfrak{I} = \int_{D^*} \mathfrak{L}\, dV$$

be stationary under all variations $\delta Q_{(\rho)}$, $\delta L^{\alpha}{}_{\beta\gamma}$ that vanish on the boundary of D^* and are such that

$$\delta L^{\alpha}{}_{[\beta\gamma]} = 0$$

everywhere in D^*. This axiom leads to the following system of field equations:

$$H^{\alpha\beta}{}_{,\beta} = S^{\alpha}, \tag{1.4}$$

$$I^{\alpha\beta}{}_{:\gamma} = \tfrac{2}{5}\, \delta^{(\alpha}_{\gamma}\, I^{\beta)\eta}{}_{:\eta} + P^{\alpha\beta}_{\cdot\,\cdot}, \tag{1.5}$$

$$Z^{(\rho)} \neq W^{(\rho)\eta}{}_{,\eta} = 0, \qquad \rho = 1, \ldots, n, \tag{1.6}$$

where

$$H^{\alpha\beta} = \mathfrak{L}_{,A_{\alpha\beta}}, \tag{1.7}$$

$$I^{\alpha\beta} = \mathfrak{L}_{,R_{\alpha\beta}}, \tag{1.8}$$

$$Z^{(\rho)} = \mathfrak{L}_{,Q_{(\rho)}}, \tag{1.9}$$

$$W^{(\rho)\eta} = \mathfrak{L}_{,Q_{(\rho),\eta}}, \tag{1.10}$$

$$S^{\alpha} = \tfrac{1}{10}\left(3 I^{\alpha\eta}{}_{:\eta} - 2\partial_{L^{\eta}{}_{\eta\alpha}} \mathfrak{L}\right), \tag{1.11}$$

$$P^{\alpha\beta}_{\cdot\,\cdot\,\gamma} = \tfrac{2}{5}\, \delta^{(\alpha}_{\gamma}\, \partial_{L^{\eta}{}_{\beta)\eta}} \mathfrak{L} - \partial_{L^{\gamma}{}_{\alpha\beta}} \mathfrak{L}, \tag{1.12}$$

and

$$I^{\alpha\beta}{}_{:\gamma} \stackrel{\text{def}}{=} I^{\alpha\beta}{}_{,\gamma} + L^{\alpha}{}_{\rho\gamma}\, I^{\rho\beta} + L^{\beta}{}_{\rho\gamma}\, I^{\alpha\rho} - L^{\rho}{}_{\rho\gamma}\, I^{\alpha\beta}. \tag{1.13}$$

From the definitions of the various terms given above, we have the additional relations

$$L^{\alpha}{}_{[\beta\gamma]} = 0, \tag{1.14}$$

$$A_{(\alpha\beta)} = 0, \quad A_{[\alpha\beta,\gamma]} = 0, \tag{1.15}$$

$$R_{[\alpha\beta]} = 0, \tag{1.16}$$

$$H^{(\alpha\beta)} = 0, \tag{1.17}$$

$$I^{[\alpha\beta]} = 0, \tag{1.18}$$

$$P^{[\alpha\beta]}_{\cdot\cdot\gamma} = 0, \tag{1.19}$$

$$P^{\alpha\eta}_{\cdot\cdot\eta} = P^{\eta\alpha}_{\cdot\cdot\eta} = 0, \tag{1.20}$$

and

$$S^{\alpha}{}_{,\alpha} = 0. \tag{1.21}$$

The system of equations (1.4)–(1.21) constitutes the variant field theory, from which we must *derive* the various physical systems by appropriate constitutive assumptions and definitions.

In order to discuss macroscopic matter in a simple and direct manner, it is convenient to split the Lagrangian function into two parts, the field Lagrangian ${}_F\mathfrak{L}$ and the matter Lagrangian ${}_M\mathfrak{L}$. The action of matter,

$${}_M\mathfrak{I} = \int_{D^*} {}_M\mathfrak{L}\, dV,$$

is developed in the next section, in which the matter Lagrangian ${}_M\mathfrak{L}$ is defined in terms of $L^{\alpha}{}_{\beta\gamma}$, the symmetric coefficient tensor $h_{\alpha\beta}$ of a quadratic form $h_{\alpha\beta}\, dx^{\alpha}\, dx^{\beta}$, and additional quantities particular to the description of macroscopic matter. We identify the $h_{\alpha\beta}$ with the first ten $Q_{(\rho)}$ and consider the combined matter and field actions as the total action,

$${}_T\mathfrak{I} = {}_M\mathfrak{I} + {}_F\mathfrak{I} = \int_{D^*} ({}_M\mathfrak{L} + {}_F\mathfrak{L})\, dV = \int_{D^*} {}_T\mathfrak{L}\, dV. \tag{1.22}$$

Extremizing the total action with respect to $L^{\alpha}{}_{\beta\gamma}$ and the $h_{\alpha\beta}$ (= the first ten $Q_{(\rho)}$), we obtain the variant field equations (1.4)–(1.6) and the subsidiary equations (1.7)–(1.21), on replacing $\mathfrak{L}$ by ${}_T\mathfrak{L}$ everywhere except in (1.7) and (1.8); ${}_F\mathfrak{L}$ remains in (1.7) and (1.8) since the matter Lagrangian does not depend on $A_{\alpha\beta}$ and $R_{\alpha\beta}$. In order for ${}_T\mathfrak{I}$ to be extremal, we must add to these equations the ones that result from extremizing the total action with respect to the choice of the quantities particular to the description of matter, that is, the remaining $Q_{(\rho)}$, where the variation of each of these quantities vanishes on the boundary of D^* in conformity with the boundary requirements on the $Q_{(\rho)}$. As is evident from the assumed functional arguments of ${}_M\mathfrak{L}$, this latter extremization of ${}_M\mathfrak{I}$, together with the field equations already obtained, is equivalent to the extremization of ${}_T\mathcal{G}$.

2. Macroscopic Matter

Since, by axiom F_2, field space is affinely connected, it is filled with a system of curves that are *paths* of the connection $L^{\alpha}{}_{\beta\gamma}$.† In order for field space to admit a description of macroscopic matter in a simple manner, it is necessary to introduce a subsidiary set of curves, called *fibers*, which have certain properties in common with the world lines of general relativity.

CONSTITUTIVE ASSUMPTION 2.1. *Let field space be such that it is filled with a 3-parameter system of fibers*

$$x^{\alpha} = x^{\alpha}(u, v, w, p),$$

where p is the parameter along the fiber, $x^{\alpha}(u, v, w, 0)$ forms a 3-dimensional continuum E_m, referred to as the matter continuum, and the Jacobian determinant of the x's with respect to the (u, v, w, p) is nonzero. In addition, let there be defined a symmetric tensor $h_{\alpha\beta}(\mathbf{x})$ such that

$$ds^2 \overset{\text{def}}{=} h_{\alpha\beta}\, dx^{\alpha}\, dx^{\beta} > 0, \qquad h = \det(h_{\alpha\beta}) \neq 0, \tag{2.1}$$

where dx^{α} is computed along a fiber, that is, computed with u, v, and w held constant.

Since the 3-dimensional matter continuum E_m is specified by $p = 0$, and field space is filled with the system of fibers considered, we may take the parameters (u, v, w) as coördinates of E_m. It will turn out that these coördinates play the role of Lagrangian coördinates. For each value of (u, v, w), say

$$(\underset{0}{u}, \underset{0}{v}, \underset{0}{w}),$$

we have a function

$$ds^2(\underset{0}{u}, \underset{0}{v}, \underset{0}{w}, p),$$

say

† T. Y. Thomas: *op. cit.*, p. 6.

$$\underset{0}{ds^2}(p),$$

which is defined by (2.1) along the fiber generated by the point

$$(\underset{0}{u}, \underset{0}{v}, \underset{0}{w})$$

of E_m. Its value is given by

$$\underset{0}{ds^2}(p) = h_{\alpha\beta}(x^\gamma(\underset{0}{u}, \underset{0}{v}, \underset{0}{w}, p)) \frac{dx^\alpha(\underset{0}{u}, \underset{0}{v}, \underset{0}{w}, p)}{dp} \frac{dx^\beta(\underset{0}{u}, \underset{0}{v}, \underset{0}{w}, p)}{dp} dp^2,$$

which may be written in the abbreviated form

$$ds^2(p) = h_{\alpha\beta}\, U^\alpha\, U^\beta\, dp^2, \tag{2.2}$$

$$U^\alpha \stackrel{\text{def}}{=} \frac{dx^\alpha}{dp}, \tag{2.3}$$

all quantities being understood to be evaluated along the fiber emanating from

$$(\underset{0}{u}, \underset{0}{v}, \underset{0}{w}).$$

If we introduce the function $f(p)$, defined by

$$f^2(p) \stackrel{\text{def}}{=} h_{\alpha\beta}\, U^\alpha\, U^\beta, \tag{2.4}$$

then we can write (2.2) in the more convenient form

$$ds^2 = f^2\, dp^2, \quad \text{or} \quad \frac{ds}{dp} = f(p). \tag{2.5}$$

It is still necessary to constrain the field space further by defining certain intrinsic functions of matter.

Constitutive Assumption 2.2. *Let there be defined over the 3-dimensional space E_m two invariable distribution functions $dm \geqq 0$ and de, which will be referred to as the affine mass and charge distributions, respectively. If Ω_m is a fiber generated from a point in E_m, and D is a given domain, then the action of matter is given by*

$$_M\mathcal{I} = \int_{E_m \cap D^*} de \int_{\Omega_m \cap D^*} L^\eta{}_{\eta\alpha}\, dx^\alpha - \int_{E_m \cap D^*} dm \int_{\Omega_m \cap D^*} ds, \tag{2.6}$$

where the first term on the right-hand side is referred to as the action of electricity and the remaining term is referred to as the action of mass.†

We shall often find it necessary to represent (2.6) as a volume integral over field space. Let μ_0 be the mass-density function of E_m; then by (2.5) we have

$$dm\, ds = \mu_0\, du\, dv\, dw\, ds = \mu_0 f\, du\, dv\, dw\, dp.$$

(Since dm is a *distribution* function over E_m, and field space is, in general, not a metric space, it follows that $dm = \mu_0\, du\, dv\, dw$ when we note that a distribution function is definable in terms of a density function by the local Euclidean–Lebesgue measure.) The Jacobian matrix of the (x^α) with respect to the (u, v, w, p) is nonsingular, by Constitutive Assumption 2.1, and hence its inverse exists. Set

$$B = \left| \det \frac{\partial(u, v, w, p)}{\partial(x^1, x^2, x^3, x^4)} \right|;$$

then we have

$$\mu_0\, du\, dv\, dw\, dp = \mu_0\, B\, dV \overset{\text{def}}{=} \mu\, dV,$$

where μ is defined to be the "world" mass-density function. Substituting this into the second term on the right-hand side of (2.6) yields

$$\int_{E_m \cap D^*} dm \int_{\Omega_m \cap D^*} ds = \int_{D^*} \mu f\, dV.$$

Similarly, we obtain

$$\int_{E_m \cap D^*} de \int_{\Omega_m \cap D^*} L^\eta{}_{\eta\alpha}\, dx^\alpha = \int_{E_m \cap D^*} de \int_{\Omega_m \cap D^*} L^\eta{}_{\eta\alpha}\, U^\alpha\, dp$$

$$= \int_{(E_m \times \Omega_m)} \rho_0\, L^\eta{}_{\eta\alpha}\, U^\alpha\, du\, dv\, dw\, dp = \int_{D^*} \rho_0\, B\, L^\eta{}_{\eta\alpha}\, U^\alpha\, dV$$

$$= \int_{D^*} \rho\, L^\eta{}_\alpha\, U^\alpha\, dV,$$

† Compare with the forms given by L. Landau and E. Lifshitz: *The Classical Theory of Fields*, Addison-Wesley Publishing Company, Inc., Cambridge, Mass. (1951), in particular pp. 272–274. It is to be noted that $L^\eta{}_{\eta\alpha}$ plays the same role here as is played by the vector potential of the electromagnetic field in the usual formulations; see the discussion following (III-3.5).

where $\rho = \rho_0 B$ is the "world" charge-density function and ρ_0 is the charge-density function in E_m. Thus we have

$$_M\mathcal{G} = \int_{D^*} \{\rho\, U^\alpha\, L^\eta{}_{\eta\alpha} - \mu\, f\}\, dV, \tag{2.7}$$

and hence the matter Lagrangian function is given by

$$_M\mathfrak{L} = \rho\, U^\alpha\, L^\eta{}_{\eta\alpha} - \mu\, f. \tag{2.8}$$

3. The Fiber Equations

In order to obtain the equations of motion of the matter continuum (equations describing the evolution of the 3-continuum E_m with the parameter p) and to extremize the total action with respect to the remaining functions $Q_{(\rho)}$ (it should be recalled that we have identified the first ten $Q_{(\rho)}$ with the independent components of the symmetric tensor $h_{\alpha\beta}$), we must extremize the functional $_M\mathcal{G}$ with respect to the choice of the fibers

$$x^\alpha = x^\alpha(u, v, w, p).$$

Denoting the operation of varying the fibers by $\underset{w}{\delta}$, we have

$$\underset{w}{\delta} \int_{E_m \cap D^*} de \int_{\Omega_m \cap D^*} L^\eta{}_{\eta\alpha}\, dx^\alpha = \int_{E_m \cap D^*} de\; \underset{w}{\delta} \int_{\Omega_m \cap D^*} L^\eta{}_{\eta\alpha}\, dx^\alpha,$$

since the fibers alone are to be varied. (The intrinsic properties of matter represented by dm and de are convected with the image of E_m, and the intrinsic properties of field space are invariant under such variations.) Now we have

$$\begin{aligned}
\underset{w}{\delta} \int_{\Omega_m \cap D^*} L^\eta{}_{\eta\alpha}\, dx^\alpha &= \int_{\Omega_m \cap D^*} (dx^\alpha\, \underset{w}{\delta}\, L^\eta{}_{\eta\alpha} + L^\eta{}_{\eta\alpha}\, d\underset{w}{\delta} x^\alpha) \\
&= \int_{\Omega_m \cap D^*} (dx^\alpha\, \underset{w}{\delta}\, L^\eta{}_{\eta\alpha} - dL^\eta{}_{\eta\alpha}\, \underset{w}{\delta} x^\alpha) \\
&\quad + \int_{\Omega_m \cap D^*} d(L^\eta{}_{\eta\alpha}\, \underset{w}{\delta} x^\alpha).
\end{aligned}$$

Thus, since $\underset{w}{\delta} x^\alpha$ vanishes at the extremities of the fibers,† we obtain

$$\underset{w}{\delta} \int_{\Omega_m \cap D^*} L^\eta{}_{\eta\alpha}\, dx^\alpha = \int_{\Omega_m \cap D^*} (L^\eta{}_{\eta\alpha,\beta} - L^\eta{}_{\eta\beta,\alpha})\, \underset{w}{\delta} x^\beta\, dx^\alpha, \tag{3.1}$$

upon noting that

$$dL^\eta{}_{\eta\alpha} = L^\eta{}_{\eta\alpha,\beta}\, dx^\beta$$

and

$$\underset{w}{\delta} L^\eta{}_{\eta\beta} = L^\eta{}_{\eta\beta,\alpha}\, \underset{w}{\delta} x^\alpha.$$

The last of these equations results from the fact that $L^\eta{}_{\eta\alpha}$ changes under $\underset{w}{\delta}$ by only that change induced in evaluating $L^\eta{}_{\eta\alpha}$ at the varied point on the fiber. Similarly, we get

$$\underset{w}{\delta} \int_{E_m \cap D^*} dm \int_{\Omega_m \cap D^*} ds = \int_{E_m \cap D^*} dm\, \underset{w}{\delta} \int_{\Omega_m \cap D^*} ds.$$

From (2.5) we have $ds = f\, dp$, so that

$$\underset{w}{\delta} \int_{\Omega_m \cap D^*} ds = \underset{w}{\delta} \int_{\Omega_m \cap D^*} f\, dp = \int_{\Omega_m \cap D^*} \underset{w}{\delta} f\, dp,$$

since p is the independent variable in the $\underset{w}{\delta}$ process and the variation vanishes at the extremity of the fiber, that is, on the intersections of the fiber with the boundary of D^*. By a straightforward application of the calculus of variations, we obtain

$$\underset{w}{\delta} \int_{\Omega_m \cap D^*} ds = \int_{\Omega_m \cap D^*} \left[f_{,x^\gamma} - \frac{d}{dp} (f_{,U^\gamma}) \right] \underset{w}{\delta} x^\gamma\, dp, \tag{3.2}$$

where $U^\alpha = dx^\alpha/dp$ by (2.3). From (2.4) we have

$$f^2 = h_{\alpha\beta}(x^\gamma)\, U^\alpha U^\beta,$$

so that

$$f_{,x^\gamma} = \frac{1}{2f} h_{\alpha\beta,\gamma}\, U^\alpha U^\beta.$$

† The initial and final states of the continuum are not to be varied, since axiom F_4 requires the vanishing of $\delta Q_{(\rho)}$ on the boundary of D^*.

Further, we have

$$f_{,U^\gamma} = \frac{1}{f} h_{\gamma\beta} U^\beta,$$

by the symmetry of $h_{\alpha\beta}$. Substituting these results back into (3.2) gives

$$\underset{w}{\delta} \int_{\Omega_m \cap D^*} ds = - \int_{\Omega_m \cap D^*} \frac{1}{2f} \Big[(h_{\gamma\alpha,\beta} + h_{\gamma\beta,\alpha} - h_{\alpha\beta,\gamma}) U^\alpha U^\beta + 2h_{\gamma\beta} \frac{dU^\beta}{dp} - 2h_{\gamma\beta} U^\beta \frac{d}{dp} (\ln f) \Big] \underset{w}{\delta x^\gamma} \, dp,$$

where the relation

$$h_{\alpha[\beta,\gamma]} U^\beta U^\gamma = 0$$

has been used. Combining (3.1) with the above result, and converting the integration over the fibers and over E_m to a volume integration over field space, we obtain

$$\underset{w}{\delta} {}_M\mathcal{I} = \int_{D^*} \Big\{ \frac{\mu}{f} \Big[h_{\gamma\beta} \frac{dU^\beta}{dp} + \frac{1}{2} (h_{\gamma\alpha,\beta} + h_{\gamma\beta,\alpha} - h_{\alpha\beta,\gamma}) U^\alpha U^\beta - h_{\gamma\beta} U^\beta \frac{d}{dp} (\ln f) \Big] + \rho (L^\eta{}_{\eta\alpha,\gamma} - L^\eta{}_{\eta\gamma,\alpha}) U^\alpha \Big\} \underset{w}{\delta x^\gamma} \, dV. \tag{3.3}$$

Noting that

$$L^\eta{}_{\eta\alpha,\delta} - L^\eta{}_{\eta\delta,\alpha} = A_{\delta\alpha}$$

and that

$$\Gamma^\alpha{}_{\beta\gamma}(h_{\mu\pi}) = \tfrac{1}{2} h^{\alpha\eta} (h_{\eta\beta,\gamma} + h_{\eta\gamma,\beta} - h_{\beta\gamma,\eta}),$$

upon annulling the variation of ${}_M\mathcal{I}$ and multiplying by $h^{\gamma\lambda}$ we get

$$\frac{\mu}{f} \Big[\frac{dU^\lambda}{dp} + \Gamma^\lambda{}_{\alpha\beta}(h_{\eta\pi}) U^\alpha U^\beta - U^\lambda \frac{d}{dp} (\ln f) \Big] + \rho \, h^{\gamma\lambda} A_{\gamma\alpha} U^\alpha = 0, \tag{3.4}$$

which are the equations governing the fibers and hence the evolution of the matter continuum E_m with p.

We shall now show that (3.4) can be reduced to a simpler form. Denote by ∇_β the covariant derivative formed from the affinity $\Gamma^\alpha{}_{\beta\gamma}(h_{\eta\pi})$. It is then evident that

$$\frac{dU^\lambda}{dp} + \Gamma^\lambda{}_{\alpha\beta}(h_{\eta\pi})\, U^\alpha\, U^\beta = U^\beta\, \nabla_\beta U^\lambda,$$

and hence (3.4) becomes

$$\frac{\mu}{f}\left\{U^\beta\, \nabla_\beta U^\lambda - U^\lambda \frac{d}{dp}(\ln f)\right\} + \rho\, h^{\gamma\lambda}\, A_{\gamma\alpha}\, U^\alpha = 0. \tag{3.5}$$

Multiplying (3.5) by $h_{\lambda\sigma}\, U^\sigma$ and noting that

$$h_{\lambda\sigma}\, U^\sigma\, U^\beta\, \nabla_\beta U^\lambda = \tfrac{1}{2}\, U^\beta\, \nabla_\beta(h_{\lambda\sigma}\, U^\lambda\, U^\sigma)$$

since

$$\nabla_\beta h_{\alpha\gamma} = 0,$$

we have

$$\frac{\mu}{f}\left\{\frac{1}{2}\, U^\beta\, \nabla_\beta(h_{\lambda\sigma}\, U^\lambda\, U^\sigma) - h_{\lambda\sigma}\, U^\gamma\, U^\sigma \frac{d}{dp}(\ln f)\right\}$$
$$+ \rho\, A_{\gamma\alpha}\, U^\gamma\, U^\alpha = 0. \tag{3.6}$$

Since the last term vanishes because of the symmetry of $U^\gamma\, U^\alpha$ and the skew-symmetry of $A_{\gamma\alpha}$ in (γ, α), and since

$$f^2 = h_{\lambda\sigma}\, U^\lambda\, U^\sigma$$

by (2.4), we have

$$\frac{1}{2}\, U^\beta\, \nabla_\beta(f^2) - f^2 \frac{d}{dp} \ln f = 0,$$

which is identically satisfied since

$$U^\beta\, \nabla_\beta f^2 = \frac{d}{dp} f^2.$$

Thus, the fiber equations (3.5) place no restriction on f, and hence we can preassign f with no loss of generality. This is equivalent to being able to preassign s with no loss of generality, as shown by (2.5). The simplest choice is $f = 1$, which is equivalent to identifying s with a linear function of p. (The choice $f = 1$ implies $ds = dp$ by (2.5), so that $s = p +$ constant.) It is also evident that $f = 1$ is a first integral of the fiber equations. We have thus proved the following result.

THEOREM 3.1. *If $\mathfrak{F}$ is a field space for which Constitutive Assumptions* 2.1 *and* 2.2 *are satisfied, then the equations of motion of the fibers in $\mathfrak{F}$ are given by*

$$\mu\, U^{\beta}\, \nabla_{\beta} U^{\lambda} + \rho\, h^{\lambda\delta}\, A_{\delta\alpha}\, U^{\alpha} = 0, \tag{3.7}$$

which admit the first integral

$$f^2 = h_{\alpha\beta}\, U^{\alpha}\, U^{\beta} = 1.$$

There are several interesting conclusions that can be drawn from (3.7).

(a) All the operations that occur in (3.7) are well defined in a subsidiary metric space with metric tensor $h_{\alpha\beta}$. (Here $h_{\alpha\beta}$ is a symmetric nonsingular tensor, by Constitutive Assumption 2.1, and ∇_{β} annuls $h_{\alpha\beta}$, so that postulates M_1, M_2, M_3 are satisfied.) Hence (3.7), although derived in field space, may be interpreted in a subsidiary metric space.

(b) If the matter continuum is such that the mass and charge distributions dm and de are nonzero only in a finite number of small disjoint regions of E_m, that is, if matter is composed of small discrete "lumps," then on averaging (3.7) over a lump, we may interpret (3.7) as the equation of motion of such a lump, where $\bar{\mu} = m_0$ is identified with the mass of the lump and $\bar{\rho} = e_0$ is identified with the charge of the lump. In this case, equations (3.7) are just those equations that describe the motion of a classic test particle with mass m_0 and charge e_0 in the presence of an electromagnetic field with field tensor $A_{\alpha\beta}$ in a metric space of the Riemannian variety.†

(c) We tentatively identified the tensor $A_{\alpha\beta}$ with the electromagnetic field tensor in III-3 on proving that $A_{\alpha\beta}$ has the operational and structural properties of an electromagnetic field tensor, that is, $A_{\alpha\beta}$ is of Maxwellian form, and corroborated this interpretation in V-5 by showing that the solution manifold of the variant field equations in a proper semiempty space with a given Lagrangian function is isomorphic with the solution manifold of vacuum electrodynamics with $A_{\alpha\beta}$ as the field tensor. Since we did not have an expression for the motion of test particles, and hence no expression for the ponderomotive force (if any) that the $A_{\alpha\beta}$ field would exert on a test particle, all that could actually be stated in V-5 was that $A_{\alpha\beta}$ is proportional to the electromagnetic field tensor

† A. Lichnerowicz: *Théories relativistes de la gravitation et de l'électromagnétisme*, Masson, Paris (1955), pp. 54–55. It is to be noted that the ρ and μ of Lichnerowicz correspond to our $\mu(-h)^{-1/2}$ and $\rho(-h)^{-1/2}$, respectively.

in vacuo. Equation (3.7) now shows that the $A_{\alpha\beta}$ field is just the electromagnetic field tensor, since the last term in that equation is the expression for the ponderomotive force generated by an electromagnetic field with field tensor $A_{\alpha\beta}$ on a test particle of charge e_0 in a physical metric space.

(d) The most important and final conclusion is obtained as follows. Let the Lagrangian function of the field be an arbitrary element of class C^2, with unspecified transformation properties under $'x = f(x)$. With this assumption, the variant field equations in general are not tensorial, and hence are not covariant with respect to the coördinate transformations $'x = f(x)$. The fiber equations, however, are tensorial, as is evident from inspection of (3.7). Thus, even if the variant field equations are non-tensorial, the interactions of the resulting fields with the matter continuum assume the covariant form (3.7) under the constitutive assumptions made previously in this chapter, and hence comply with the principle of observational covariance.

4. Explicit Form of the Field Variables

Having obtained a description of macroscopic matter by specifying the matter Lagrangian and obtaining the fiber equations, we now begin the study of the corresponding fields generated by the following assumption.

CONSTITUTIVE ASSUMPTION 4.1. *The field Lagrangian is the scalar density*†

$$_F\mathfrak{L} = \sqrt{-h}\left(\Lambda + a\, h^{\alpha\beta} R_{\alpha\beta} - \frac{b}{4} A_{\alpha\beta} A_{\gamma\lambda} h^{\alpha\gamma} h^{\beta\lambda}\right), \tag{4.1}$$

where a and b are positive constants and $h_{\alpha\beta}$ is the symmetric tensor introduced in Constitutive Assumption 2.1 *and identified with the first ten $Q_{(\rho)}$ variables, so that the relations*

$$h = \det(h_{\alpha\beta}) < 0 \tag{4.2}$$

and

$$h\, h^{\alpha\beta} = h_{,h_{\alpha\beta}} \tag{4.3}$$

hold.

† Compare with J. G. Fletcher: "Local Conservation Laws in Generally Covariant Theories," *Rev. Mod. Phys.*, **32** (1960), pp. 65–87, Eq. 2.10a and 2.10d. It should be noted that in Fletcher's formulations, as well as in the formulations of most others, the $A_{\alpha\beta}$ field is a subsidiary field defined over a Riemannian metric space with metric tensor $h_{\alpha\beta}$, but in our formulation the $A_{\alpha\beta}$ is an intrinsic geometric tensor representing curvature in field space, and $h_{\alpha\beta}$ is a subsidiary symmetric tensor field defined over field space, which is not a metric space unless $A_{\alpha\beta} = 0$.

The choice (4.1) is not as arbitrary as might appear on first examination. To demonstrate this, we first write (4.1) in the equivalent form

$$_F\mathfrak{L} = \frac{\det\{a^{-1/4}(\sqrt{a\,R}\,h_{\alpha\beta} + i\sqrt{b/2}\,A_{\alpha\beta})\} - \det\{i\,a^{-1/4}\sqrt{b/2}\,A_{\alpha\beta}\}}{R\sqrt{-h}}$$

where

$$R = R_{\alpha\beta}\,h^{\alpha\beta}.$$

Thus, since $h_{\alpha\beta}$ and $A_{\alpha\beta}$ will be shown to be proportional to the gravitational field potentials and the electromagnetic field tensor, we see that $_F\mathfrak{L}$ is proportional to the determinant of the hermitian field $a^{-1/4}\sqrt{a\,R}\,h_{\alpha\beta} + i\sqrt{b/2}\,A_{\alpha\beta}$ of gravitation and electromagnetism diminished by the determinant of the hermitian field $i\,a^{-1/4}\sqrt{b/2}\,A_{\alpha\beta}$ of electromagnetism.

The total Lagrangian

$$_T\mathfrak{L} = {}_M\mathfrak{L} + {}_F\mathfrak{L},$$

is obtained by combining (4.1) and (2.8):

$$_T\mathfrak{L} = \sqrt{-h}\left\{\Lambda + a\,h^{\alpha\beta}\,R_{\alpha\beta} - \frac{b}{4}\,A_{\alpha\beta}\,A_{\gamma\lambda}\,h^{\alpha\gamma}\,h^{\beta\lambda}\right\} - \mu f + \rho\,U^{\alpha}\,L^{\eta}{}_{\eta\alpha}. \tag{4.4}$$

It is evident from inspection of (4.4) that $_T\mathfrak{L}$ is not invariant, because of the term $\rho\,U^{\alpha}\,L^{\eta}{}_{\eta\alpha}$. *This noninvariance of $_T\mathfrak{L}$ gives us no cause for concern under the variant field theory*, although those accustomed to invariant formulations would probably try to eliminate such a "thorn in the side." Such eliminations are usually accomplished by subtracting from the Lagrangian function a total divergence, say $K^{\alpha}{}_{,\alpha}$, since such subtraction would leave the field equations unchanged. The prospects of such an elimination will be considered at the close of this section.

With the total Lagrangian, we may explicitly evaluate the field variables $H^{\alpha\beta}$, $I^{\alpha\beta}$, S^{α}, $P^{\alpha\beta}{}_{\cdot\cdot\gamma}$, $Z^{(\rho)}$, and $W^{(\rho)\eta}$ in terms of the fundamental variables $A_{\alpha\beta}$ and $R_{\alpha\beta}$ and the subsidiary variables $h_{\alpha\beta}$, U^{α}, ρ, and μ. Since we have identified the $h_{\alpha\beta}$ with the first ten $Q_{(\rho)}$ and have obtained the fiber equations corresponding to the remaining $Q_{(\rho)}$, we shall write $Z^{\alpha\beta}$ and $W^{\alpha\beta\eta}$ for the expressions

$$\mathfrak{L}_{,h_{\alpha\beta}} \quad\text{and}\quad \mathfrak{L}_{,h_{\alpha\beta,\eta}},$$

respectively. The tensors $h_{\alpha\beta}$ and $h^{\alpha\beta}$ will be used to raise and lower indices in the usual manner, although it must be remembered that covariant differentiation and the raising and lowering of indices do not commute unless the components of affine connection used in the covariant differentiation process are the Christoffel symbols formed from $h_{\alpha\beta}$. Substituting (4.4) into (1.7)–(1.10) yields

$$H^{\alpha\beta} = -\frac{b}{2}\sqrt{-h}\,A^{\alpha\beta}_{\cdot\cdot}, \tag{4.5}$$

$$I^{\alpha\beta} = a\sqrt{-h}\,h^{\alpha\beta}, \tag{4.6}$$

$$Z^{\alpha\beta} = -a\sqrt{-h}\,(R^{\alpha\beta}_{\cdot\cdot} - \tfrac{1}{2}R\,h^{\alpha\beta}) + \frac{\Lambda}{2}\sqrt{-h}\,h^{\alpha\beta}$$

$$-\frac{\mu}{2}\frac{U^\alpha U^\beta}{f} + \frac{b}{2}(A^{\alpha\eta}_{\cdot\cdot}A^{\beta}_{\cdot\eta} - \tfrac{1}{4}h^{\alpha\beta}A^{\lambda\eta}_{\cdot\cdot}A_{\lambda\eta})\sqrt{-h}, \tag{4.7}$$

and

$$W^{\alpha\beta\eta} \equiv 0, \tag{4.8}$$

where

$$R = R_{\alpha\beta}\,h^{\alpha\beta}, \qquad A^{\alpha\beta}_{\cdot\cdot} = A_{\eta\lambda}\,h^{\eta\alpha}\,h^{\lambda\beta},$$

and so forth. The term $\mu\,U^\alpha\,U^\beta/2f$ is obtained as follows. We have

$$(\mu f)_{,h_{\alpha\beta}} = \mu f_{,h_{\alpha\beta}}$$

and

$$2f f_{,h_{\alpha\beta}} = U^\alpha\,U^\beta,$$

since

$$f^2 = h_{\alpha\beta}\,U^\alpha\,U^\beta.$$

Thus we get

$$(\mu f)_{,h_{\alpha\beta}} = \frac{\mu}{2f}U^\alpha\,U^\beta.$$

By a direct calculation, we have

$$\partial_{L^\alpha{}_{\beta\gamma}}\,T\mathfrak{L} = \rho\,U^{(\beta}\,\delta^{\gamma)}_\alpha, \tag{4.9}$$

so that

$$\partial_{L^\eta{}_{\eta\alpha}}\,T\mathfrak{L} = \tfrac{5}{2}\rho\,U^\alpha. \tag{4.10}$$

Substituting (4.9) and (4.10) into (1.12) gives

$$P^{\alpha\beta}_{\cdot\cdot\gamma} = 0. \tag{4.11}$$

Finally, substituting (4.5) and (4.10) into (1.11) yields

$$S^{\alpha} = \frac{3}{10}\, a(\sqrt{-h}\, h^{\alpha\beta})_{;\beta} - \frac{\rho}{2}\, U^{\alpha}. \tag{4.12}$$

Equations (4.5)–(4.8), together with (4.11) and (4.12), give the variables required for the evaluation of the variant field. Inspection of (4.5) shows that $I^{\alpha\beta}$ is a tensor density, and hence the colon derivative is the covariant derivative. From the tensor-density character of $I^{\alpha\beta}$, it follows that all the remaining field variables are also tensor densities. Under the axioms of field space, these tensor densities must satisfy the variant field equations (1.4)–(1.6). Substitution of the evaluated field variables into these equations yields

$$(-\tfrac{1}{2}\, b\, \sqrt{-h}\, A^{\alpha\beta}_{\cdot\cdot})_{,\beta} = \frac{3}{10}\, (a\, \sqrt{-h}\, h^{\alpha\beta})_{;\beta} - \frac{\rho}{2}\, U^{\alpha}, \tag{4.13}$$

$$(a\, \sqrt{-h}\, h^{\alpha\beta})_{;\gamma} = \tfrac{2}{5}\, \delta^{(\alpha}_{\gamma}\, (a\, \sqrt{-h}\, h^{\beta)\eta})_{;\eta}, \tag{4.14}$$

$$-a\, \sqrt{-h}\, (R^{\alpha\beta}_{\cdot\cdot} - \tfrac{1}{2}\, R\, h^{\alpha\beta}) + \frac{\Lambda}{2}\, \sqrt{-h}\, h^{\alpha\beta} - \frac{\mu}{2f}\, U^{\alpha}\, U^{\beta}$$
$$+ \frac{b}{2}\, (A^{\alpha\eta}_{\cdot\cdot}\, A^{\beta}_{\cdot\eta} - \tfrac{1}{4}\, h^{\alpha\beta}\, A^{\gamma\eta}_{\cdot\cdot}\, A_{\gamma\eta})\, \sqrt{-h} = 0. \tag{4.15}$$

We have thus proved the following result.

THEOREM 4.1. *Under the axioms of field space the set of nonidentically vanishing field variables $I^{\alpha\beta}$, $H^{\alpha\beta}$, S^{α}, and $Z^{\alpha\beta}$, obtained from the Lagrangian function* (4.4), *are tensor densities governed by the system of equations* (4.13)–(4.15), *and the field U^{α} is governed by the system* (3.7).

The continuation from this point requires the results of Chapters IV and V. In order to apply these results, we must first determine whether the field $I^{\alpha\beta}$ is a member of the nonnull class, the seminull class, the proper null class, or the improper null class.

From (4.5), we have

$$I \stackrel{\text{def}}{=} \det(I^{\alpha\beta}) = a^4\, h. \tag{4.16}$$

Since $h \neq 0$ under Constitutive Assumption 2.1, we have by (4.6) and

(4.16) either $I^{\alpha\beta} = 0$ or $I \neq 0$ as $a = 0$ or $a \neq 0$. Hence $I^{\alpha\beta}$ is never a member of the improper null class. We thus have to consider three cases.

Case 1: $a \neq 0$, $I^{\alpha\eta}{}_{:\eta} \neq 0$. For this case, $I^{\alpha\beta}$ is a member of the nonnull class, so that the results of Chapter IV are applicable. In addition, by (4.5), $I^{\alpha\beta}$ is a tensor density having a form identical with that given in (IV-5.1). All the results of IV-5 therefore apply directly to this case. The field equations (4.14) are thus equivalent to the system of equations

$$h_{\alpha\beta;\gamma} = M_\gamma h_{\alpha\beta} - 2h_{\gamma(\alpha} M_{\beta)}, \tag{4.17}$$

where M_γ is an arbitrary nonzero vector, so that *in this case field space is well based with base field* $h_{\alpha\beta}$ (see I-5). By means of the results of Chapter IV, equation (4.14) is readily solvable for $L^\alpha{}_{\beta\gamma}$, yielding

$$L^\alpha{}_{\beta\gamma} = \Gamma^\alpha{}_{\beta\gamma}(h_{\eta\pi}) - \tfrac{1}{2}\,(M_\gamma\,\delta^\alpha_\beta + M_\beta\,\delta^\alpha_\gamma - 3M^\alpha_{\cdot}\,h_{\beta\gamma}) \tag{4.18}$$

as the general form for the affinity. By (IV-5.9) and (IV-5.10), we also have

$$R_{\mu\lambda} = J_{\mu\lambda} - \tfrac{3}{2}\,h_{\mu\lambda}\,\nabla_\eta M^\eta_{\cdot} + \tfrac{3}{2}\,M_\mu\,M_\lambda \tag{4.19}$$

and

$$A_{\mu\lambda} = 2M_{[\mu,\lambda]}, \tag{4.20}$$

where $J_{\mu\lambda}$ is the contraction of the curvature tensor formed from the affinity $\Gamma^\alpha{}_{\beta\gamma}(h_{\eta\pi})$, and ∇_γ stands for the covariant derivative formed from $\Gamma^\alpha{}_{\beta\gamma}(h_{\eta\pi})$. Thus, by (III-3.5) and (4.20), M_α is related to the vector potential f_α of the Maxwellian field by the equation

$$M_\alpha = \tfrac{1}{2}\,f_\alpha. \tag{4.21}$$

We can use (4.21) to rewrite the above equations in terms of the vector potential f_α, in a form that we shall later require:

$$h_{\alpha\beta;\gamma} = \tfrac{1}{2}\,h_{\alpha\beta}\,f_\gamma - h_{\gamma(\alpha}\,f_{\beta)}, \tag{4.22}$$

$$L^\alpha{}_{\beta\gamma} = \Gamma^\alpha{}_{\beta\gamma}(h_{\eta\pi}) - \tfrac{1}{4}\,(f_\gamma\,\delta^\alpha_\beta + f_\beta\,\delta^\alpha_\gamma - 3f^\alpha_{\cdot}\,h_{\beta\gamma}), \tag{4.23}$$

$$R_{\eta\lambda} = J_{\eta\lambda} - \tfrac{3}{4}\,h_{\eta\lambda}\,\nabla_\rho f^\rho_{\cdot} + \tfrac{3}{8}\,f_\eta\,f_\lambda, \tag{4.24}$$

$$A_{\mu\lambda} = f_{[\mu,\lambda]}. \tag{4.25}$$

Summarizing, we have the following result.

THEOREM 4.2. *Let* $\mathfrak{F}$ *be a field space in which Constitutive Assumptions* 2.1, 2.2, *and* 4.1 *are satisfied. In the case*

$$a \neq 0, \qquad I^{\alpha\eta}{}_{:\eta} \neq 0,$$

$\mathfrak{F}$ *is a well-based space with base field* $h_{\alpha\beta}$; *the affine connection is given by* (4.23) *in terms of the 14 potential functions* $h_{\alpha\beta}$ *and* f_γ, *which must satisfy the 14 field equations* (4.13) *and* (4.15); *and the field equations* (4.14) *are identically and algebraically satisfied by this* $L^\alpha{}_{\beta\gamma}$. *In addition, equations* (4.22), (4.24), *and* (4.25) *hold.*

Case 2: $a \neq 0$, $I^{\alpha\eta}{}_{:\eta} = 0$. By Theorem 4.1, $I^{\alpha\beta}$ is a tensor density, and we have

$$P^{\alpha\beta}_{\cdot\cdot\gamma} = 0$$

by (4.11). Since $a \neq 0$ implies $I \neq 0$, the hypotheses of Theorem V-3.4′ are satisfied. We thus have established the following result.

THEOREM 4.3. *Let* $\mathfrak{F}$ *be a field space in which Constitutive Assumptions* 2.1, 2.2, *and* 4.1 *are satisfied. In the case*

$$a \neq 0, \quad I^{\alpha\eta}{}_{:\eta} = 0,$$

$\mathfrak{F}$ *is a metric space with metric tensor* $h_{\alpha\beta}$, *and* $L^\alpha{}_{\beta\gamma} = \Gamma^\alpha{}_{\beta\gamma}(h_{\eta\pi})$.

Case 3: $a = 0$. For this case, $I^{\alpha\beta}$ is a member of the proper null class, and in addition we lose the field equations (4.14) since they are identically satisfied. Equation (4.13) thus becomes

$$(-\tfrac{1}{2}\, b \sqrt{-h}\, A^{\alpha\beta}_{\cdot\cdot})_{,\beta} = -\frac{\rho}{2}\, U^\alpha, \tag{4.26}$$

so that we must require the field tensor $A_{\alpha\beta}$ to be nonzero for some (α, β) if the field equations are not to be identically satisfied. If $A_{\alpha\beta} \neq 0$ for some (α, β), then by the definition of $A_{\alpha\beta}$ we must have

$$L^\eta{}_{\eta[\beta,\alpha]} \neq 0.$$

This condition is easily satisfied if we take (4.23) as the definition of $L^\alpha{}_{\beta\gamma}$ provided f_α is not a gradient vector. The choice is not unique, however; we could equally well have taken (V-5.6), again requiring f_α not to be a gradient vector, and have satisfied the requirement $A_{\alpha\beta} \neq 0$. If we are to obtain a deterministic system of field equations in the light of this indeterminism in the form of the affinity, it can be shown that the tensor field $h_{\alpha\beta}$ must be assumed to be a given field and thus not subject to variation. Hence the field equations (4.15) no longer hold and we are left with the equation (4.26) only. We thus have the following result.

THEOREM 4.4. *Let* $\mathfrak{F}$ *be a field space for which Constitutive Assumptions* 2.1, 2.2, *and* 4.1 *are satisfied. In the case* $a = 0$, *the field equations reduce to* $0 = 0$ *unless*

$$L^{\eta}{}_{\eta[\beta,\alpha]} \neq 0 \tag{4.27}$$

for some (α, β). *If* (4.27) *holds, then the choice of the affinity is not unique and this nonuniqueness requires that the* $h_{\alpha\beta}$ *field be an assigned field. If* f_α *is not a gradient vector, then either* (4.23) *or* (V-5.6) *can be used as the potential forms for the affinity and the only surviving field equations are* (4.26).

We now have the results required to examine the possible elimination of the noninvariance of ${}_T\mathfrak{L}$ associated with $\rho U^\alpha L^{\eta}{}_{\eta\alpha}$. From (4.23), we have

$$L^{\epsilon}{}_{\epsilon\alpha} = \Gamma^{\epsilon}{}_{\epsilon\alpha}(h_{\xi\zeta}) - \tfrac{1}{2} f_\alpha = (\ln\sqrt{-h})_{,\alpha} - \tfrac{1}{2} f_\alpha,$$

since

$$\Gamma^{\epsilon}{}_{\epsilon\alpha}(h_{\xi\zeta}) = (\ln\sqrt{-h})_{,\alpha}.$$

Thus we obtain

$$\rho\, U^\alpha L^{\epsilon}{}_{\epsilon\alpha} = -\tfrac{1}{2}\rho\, U^\alpha f_\alpha + \rho\, U^\alpha (\ln\sqrt{-h})_{,\alpha}.$$

Now, substituting equations (4.6) and (4.12) into $S^\alpha{}_{,\alpha} = 0$, we have

$$(\rho\, U^\alpha)_{,\alpha} = \tfrac{3}{5}(I^{\alpha\epsilon}{}_{;\epsilon})_{,\alpha}.$$

The term $I^{\alpha\epsilon}{}_{;\epsilon}$ may be directly evaluated using (4.12) and (4.22), giving

$$I^{\alpha\epsilon}{}_{;\epsilon} = \tfrac{5}{2} a \sqrt{-h} f^\alpha, \tag{4.28}$$

so that

$$(\rho\, U^\alpha)_{,\alpha} = \tfrac{3}{2} a (\sqrt{-h} f^\alpha)_{,\alpha}.$$

Thus since

$$(\ln\sqrt{-h})_{,\alpha}\, \rho\, U^\alpha = (\rho\, U^\alpha \ln\sqrt{-h})_{,\alpha} - (\ln\sqrt{-h})(\rho\, U^\alpha)_{,\alpha},$$

we have

$$\rho\, U^\alpha L^\epsilon{}_{\epsilon\alpha} = -\tfrac{1}{2}\,\rho\, U^\alpha f_\alpha - \tfrac{3}{2}\, a\, (\sqrt{-h}\, f^{\alpha}_{\cdot})_{,\alpha} \ln \sqrt{-h} + (\rho\, U^\alpha \ln \sqrt{-h})_{,\alpha}.$$

Hence, if we subtract the divergence term $(\rho\, U^\alpha \ln \sqrt{-h})_{,\alpha}$ from ${}_T\mathfrak{L}$, the last term of ${}_T\mathfrak{L}$ becomes

$$-\tfrac{1}{2}\,\rho\, U^\alpha f_\alpha - \tfrac{3}{2}\, a\, (\sqrt{-h}\, f^{\alpha}_{\cdot})_{,\alpha} \ln \sqrt{-h}.$$

We thus see that only if either

(i) $a = 0$

or

(ii) the Lorentz gauge condition

$$(\sqrt{-h}\, f^{\alpha}_{\cdot})_{,\alpha} = 0$$

is satisfied does the last term of ${}_T\mathfrak{L}$ become the scalar density that is usually encountered in variational representations of electrodynamics, namely

$$-\tfrac{1}{2}\,\rho\, U^\alpha f_\alpha.$$

5. Maxwell–Lorentz Electrodynamics

We shall show that the classical Maxwell–Lorentz electrodynamics is obtained as a special case of the variant field theory. Set $a = 0$ and $b = 1$ in (4.4). The $I^{\alpha\beta}$ field then vanishes, so that the results of Case 3 of Section 4 are applicable. Hence we obtain the system of equations

$$(\sqrt{-h}\, A^{\alpha\beta}_{\cdot\cdot})_{,\beta} = +\rho\, U^\alpha \tag{5.1}$$

and

$$\mu\, U^\beta\, \nabla_\beta U^\lambda + \rho\, h^{\lambda\delta}\, A_{\delta\nu}\, U^\nu = 0 \tag{5.2}$$

for the determination of the 8 unknown functions U^α and f_α, where

$$A_{\alpha\beta} = f_{[\alpha,\beta]}, \tag{5.3}$$

$$S^\alpha = -\tfrac{1}{2}\,\rho\, U^\alpha, \qquad S^\alpha{}_{,\alpha} = 0, \tag{5.4}$$

$$L^{\alpha}{}_{\beta\gamma} = \Gamma^{\alpha}{}_{\beta\gamma}(h_{\xi\zeta}) - \tfrac{1}{4}\,[f_{\gamma}\,\delta^{\alpha}_{\beta} + f_{\beta}\,\delta^{\alpha}_{\gamma} - 3f^{\alpha}_{\cdot}\,h_{\beta\gamma}], \tag{5.5}$$

and

$$h_{\alpha\beta;\gamma} = \tfrac{1}{2}\,f_{\gamma}\,h_{\alpha\beta} - h_{\gamma(\alpha}\,f_{\beta)}. \tag{5.6}$$

As noted in Section 4, the tensor $h_{\alpha\beta}$ is assumed given in this case. Since $h_{\alpha\beta}$ is the coefficient tensor of the fundamental quadratic form, the ability to choose $h_{\alpha\beta}$ is equivalent to the ability to choose the coördinate system in which the above equations are to be solved. (Equation (5.6) does not place any restriction on the tensor $h_{\alpha\beta}$ other than that of differentiability, since (5.5) implies that (5.6) is identically satisfied for arbitrary $h_{\alpha\beta}$.) The Lorentz gauge condition

$$(\sqrt{-h}\,f^{\alpha}_{\cdot})_{,\alpha} = 0$$

can be satisfied with no loss of generality, since the vector f_{α} is determined by (5.1) and (5.3) only to within an arbitrary gradient vector (to within a gauge transformation) and the field $A_{\alpha\beta}$, together with the Lagrangian function in this case, is invariant under such gauge transformations. Since $\sqrt{-h}\,A^{\alpha\beta}_{\cdot\cdot}$ is a skew-symmetric tensor density and $\rho\,U^{\alpha}$ is a vector density, the above system of field equations can be written as

$$\nabla_{\beta}(\sqrt{-h}\,A^{\alpha\beta}_{\cdot\cdot}) = \rho\,U^{\alpha} = -2S^{\alpha}, \tag{5.7}$$

$$A_{\alpha\beta} = \nabla_{[\beta}f_{\alpha]}, \tag{5.8}$$

$$\nabla_{\alpha}S^{\alpha} = 0, \qquad \nabla_{\alpha}(\sqrt{-h}\,f^{\alpha}_{\cdot}) = 0, \tag{5.9}$$

$$\mu\,U^{\beta}\,\nabla_{\beta}U^{\alpha} + \rho\,A^{\alpha}_{\cdot\beta}\,U^{\beta} = 0, \tag{5.10}$$

where ∇_{β} stands for the covariant derivative formed from the affinity $\Gamma^{\alpha}{}_{\beta\gamma}(h_{\eta\pi})$. Since

$$\nabla_{\gamma}h_{\alpha\beta} = 0,$$

this system of field equations, although derived in the present nonmetric abstract field space, assumes a form that is well defined in a *subsidiary* physical metric space of the Riemannian variety with metric tensor $h_{\alpha\beta}$, and is just the system of equations that describes Maxwell–Lorentz electrodynamics.† We have thus proved the following result.

† A. Lichnerowicz: "Sur les équations relativistes de l'électromagnétisme," *Ann. Sci. Ecole Norm. Sup.*, **60** (1943), pp. 247–288.

THEOREM 5.1. *If $\mathfrak{F}$ is a field space in which Constitutive Assumptions* 2.1, 2.2, *and* 4.1 *are satisfied, and if $a = 0$, $b = 1$, then the variant field equations yield a system of equations in which only operations that are well defined in a subsidiary physical metric space with given metric tensor $h_{\alpha\beta}$ occur, and this system of equations is identical to the Maxwell–Lorentz equations for the electromagnetic field with field tensor $A_{\alpha\beta}$ and field potential f_α.*

It is of interest to note that the field $A_{\alpha\beta}$, which is a contraction of the curvature tensor in the abstract field space, assumes the role of the electromagnetic field tensor in physical metric space. The geometry of the abstract field space and the physical metric space are quite dissimilar, however, and the two spaces are distinct. If this were not so, the field $A_{\alpha\beta}$ would vanish identically if field space were also required to be a metric space, as shown in Chapter II. *Curvature expressions in the abstract field space are thus not necessarily the same thing as curvature expressions in the physical metric space of our experience. In general, the abstract curvature expressions become geometric representations of the various physical fields of force or their derivatives.*

Examining the governing equations of this case (equations (5.7)–(5.10)) we see that there is no statement made concerning the structure of the $R_{\alpha\beta}$ field. This field, however, is uniquely determined: Since $h_{\alpha\beta}$ is assumed given, we see by (5.5) and (5.6) that the abstract field space is well based, with base field $h_{\alpha\beta}$; thus, using (4.24), we have

$$R_{\alpha\beta} = J_{\alpha\beta} - \tfrac{3}{4}\, h_{\alpha\beta}\, \nabla_\rho f^\rho_{\cdot} + \tfrac{3}{8}\, f_\alpha f_\beta,$$

which by $(5.9)_2$ reduces to

$$R_{\alpha\beta} = J_{\alpha\beta} + \tfrac{3}{8}\, f_\alpha f_\beta. \tag{5.11}$$

The important result to note is that by use of abstract field space we have obtained a *purely geometric* representation of the electromagnetic field theory.

6. Gravitation in the Absence of Electromagnetism

We shall show that gravitation in the absence of electromagnetism is a special case of the variant field theory. Set $b = \rho = 0$ in (4.4) and assume $a \neq 0$. Under these assumptions, (4.13) implies

$$\tfrac{3}{10}\,(a\sqrt{-h}\, h^{\alpha\beta})_{;\beta} = \tfrac{3}{10}\, I^{\alpha\beta}{}_{;\beta} = 0.$$

Substituting this result into (4.14) and noting that $P^{\alpha\beta}_{\cdot\cdot\gamma} = 0$ (compare (4.11)), we obtain

$$J^{\alpha\beta}{}_{;\gamma} = 0.$$

The hypotheses of Theorem 4.3 are thus satisfied, and hence field space is a metric space with metric tensor $h_{\alpha\beta}$ and affine connection $\Gamma^{\alpha}{}_{\beta\gamma}(h_{\eta\pi})$. The Maxwellian field tensor $A_{\alpha\beta}$ thus vanishes by Theorem II-2.2, so that the conditions $b = \rho = 0$ and $a \neq 0$ imply the absence of electromagnetic effects. (By Theorem IV-4.3, the equations $0 = A_{\alpha\beta} = f_{[\alpha,\beta]}$ imply that we can take $f_\alpha = 0$ with no loss of generality.) The only surviving field equations are thus the 10 equations

$$a\left(J^{\alpha\beta}_{\cdot\cdot} - \frac{1}{2}\, h^{\alpha\beta}\, J_{\gamma\lambda}\, h^{\gamma\lambda}\right) - \frac{\Lambda}{2}\, h^{\alpha\beta} = -\left(\frac{\mu}{2}\sqrt{-h}\right) U^\alpha\, U^\beta, \tag{6.1}$$

where $J_{\alpha\beta}$ is the curvature tensor formed from $\Gamma^{\alpha}{}_{\beta\gamma}(h_{\eta\pi})$ (that is, $J_{\alpha\beta} = R_{\alpha\beta}$ in a metric space), and the 4 equations

$$U^\beta\, U^\alpha{}_{;\beta} = 0, \tag{6.2}$$

for the determination of the 14 unknowns $h_{\alpha\beta}$ and U^α. Equations (6.1) and (6.2) are, however, just those equations that describe Einsteinian gravitation theory in the absence of electromagnetic effects, in which $-\Lambda/2a$ is identified with the cosmological constant and $1/a$ with the matter-field interaction coefficient. We have thus proved the following result.

THEOREM 6.1. *Let* $\mathfrak{F}$ *be a field space for which Constitutive Assumptions* 2.1, 2.2, *and* 4.1 *are satisfied. If* $b = \rho = 0$, *then* $\mathfrak{F}$ *is a metric space with metric tensor* $h_{\alpha\beta}$, *and the variant field equations reduce to the Einstein gravitation equations* (6.1) *and* (6.2), *with* $-\Lambda/2a$ *as the cosmological constant and* $1/a$ *as the matter-field interaction coefficient.*†

Equations (6.1) and (6.2) imply that the mass-density μ is conserved. Set

$$G^{\alpha\beta} = J^{\alpha\beta}_{\cdot\cdot} - \tfrac{1}{2}\, h^{\alpha\beta}\, J_{\gamma\lambda}\, h^{\gamma\lambda}; \tag{6.3}$$

then by (II-2.8) we have

† It is to be noted that this result has been established without stating a principle of equivalence. For similar results, see R. H. Dicke: "Gravitation without a Principle of Equivalence," *Rev. Mod. Phys.*, **29** (1957), pp. 363–376.

$$G^{\alpha\beta}{}_{;\beta} = 0 \tag{6.4}$$

since covariant differentiation and the raising of indices commute in a metric space. Substituting (6.1) into (6.3) yields

$$a\, G^{\alpha\beta} = \frac{\Lambda}{2} h^{\alpha\beta} - \frac{\mu}{2} (-h)^{-1/2}\, U^\alpha\, U^\beta;$$

further, we have $h^{\alpha\beta}{}_{;\gamma} = 0$, since field space is a metric space with metric tensor $h_{\alpha\beta}$ in this case. Hence, from (6.4) we obtain

$$\left(\frac{\mu}{2} (-h)^{-1/2}\, U^\alpha\, U^\beta\right)_{;\beta} = 0. \tag{6.5}$$

Expanding (6.5) gives

$$\mu\, (-h)^{-1/2}\, U^\beta\, U^\alpha{}_{;\beta} + U^\alpha\, (\mu\, (-h)^{-1/2}\, U^\beta)_{;\beta} = 0,$$

which by (6.2) reduces to

$$(\mu\, (-h)^{-1/2}\, U^\beta)_{;\beta} = 0. \tag{6.6}$$

But we have $h_{;\gamma} = 0$, so that (6.6) is equivalent to

$$(\mu\, U^\beta)_{;\beta} = 0.$$

Since $\mu\, U^\beta$ is a vector density, it follows that

$$(\mu\, U^\beta)_{;\beta} = (\mu\, U^\beta)_{,\beta},$$

and hence we finally obtain the conservation law

$$(\mu\, U^\beta)_{,\beta} = 0.$$

Chapter VII

Interacting Fields

IT WAS POINTED OUT in VI-4 that under Constitutive Assumptions VI-2.1, VI-2.2, and VI-4.1, there were three cases to be considered. The second case,

$$a \neq 0, \qquad I^{\alpha\eta}{}_{:\eta} = 0,$$

was shown to yield the Einstein gravitation theory in the absence of electromagnetic effects; and the third case,

$$a = 0,$$

was shown to yield Maxwell–Lorentz electrodynamics. In both of these cases, the fundamental field variables were of geometric origin. The pure gravitational problem and the pure electromagnetic problem were thus represented from a common geometric viewpoint in terms of curvature expressions of the abstract field space and subsidiary vector and tensor fields defined over that space. The first case,

$$a \neq 0, \qquad I^{\alpha\eta}{}_{:\eta} \neq 0,$$

is the subject of this chapter. It will be shown that this case represents interacting gravitational and electromagnetic fields from a common geometric basis and yields a set of results of no mean significance in their implications of the structure of matter.

1. The Field Equations

We first summarize the salient results established in Chapter VI. Under Constitutive Assumptions VI-2.1, VI-2.2, and VI-4.1, the total Lagrangian is given by

$$_T\mathfrak{L} = \sqrt{-h}\left\{\Lambda + a\, h^{\alpha\beta} R_{\alpha\beta} - \frac{b}{4} A_{\alpha\beta} A_{\gamma\lambda} h^{\alpha\gamma} h^{\beta\lambda}\right\}$$

$$+ \rho\, U^\alpha L^\eta{}_{\eta\alpha} - \mu f, \tag{1.1}$$

and hence the nonvanishing field variables are

$$I^{\alpha\beta} = a\sqrt{-h}\, h^{\alpha\beta}, \tag{1.2}$$

$$H^{\alpha\beta} = -\frac{b}{2}\sqrt{-h}\, A^{\alpha\beta}_{\cdot\cdot}, \tag{1.3}$$

$$Z^{\alpha\beta} = -a\sqrt{-h}\left(R^{\alpha\beta}_{\cdot\cdot} - \frac{1}{2} R\, h^{\alpha\beta}\right) + \frac{\Lambda}{2}\sqrt{-h}\, h^{\alpha\beta}$$

$$-\frac{\mu}{2f} U^\alpha U^\beta + \frac{b}{2}\left(A^{\alpha\eta}_{\cdot\cdot} A^\beta_{\cdot\eta} - \frac{1}{4} h^{\alpha\beta} A^{\lambda\eta}_{\cdot\cdot} A_{\lambda\eta}\right)\sqrt{-h}, \tag{1.4}$$

$$S^\alpha = \frac{3a}{10}\left(\sqrt{-h}\, h^{\alpha\beta}\right)_{;\beta} - \frac{\rho}{2} U^\alpha. \tag{1.5}$$

Substituting these field variables into the variant field equations, so that axiom F_4 is satisfied, we obtain

$$\left(-\frac{b}{2}\sqrt{-h}\, A^{\alpha\beta}_{\cdot\cdot}\right)_{,\beta} = \frac{3}{10}\left(a\sqrt{-h}\, h^{\alpha\beta}\right)_{;\beta} - \frac{\rho}{2} U^\alpha, \tag{1.6}$$

$$\left(a\sqrt{-h}\, h^{\alpha\beta}\right)_{;\gamma} = \tfrac{2}{5}\,\delta^{(\alpha}_\gamma \left(a\sqrt{-h}\, h^{\beta)\eta}\right)_{;\eta}, \tag{1.7}$$

$$-a\sqrt{-h}\left(R^{\alpha\beta}_{\cdot\cdot} - \frac{1}{2} R\, h^{\alpha\beta}\right) + \frac{\Lambda}{2}\sqrt{-h}\, h^{\alpha\beta} - \frac{\mu}{2f} U^\alpha U^\beta$$

$$+ \frac{b}{2}\left(A^{\alpha\eta}_{\cdot\cdot} A^\beta_{\cdot\eta} - \frac{1}{4} h^{\alpha\beta} A^{\lambda\eta}_{\cdot\cdot} A_{\lambda\eta}\right)\sqrt{-h} = 0, \tag{1.8}$$

together with

$$\frac{\mu}{f}\left\{U^\beta \nabla_\beta U^\lambda - U^\lambda\, d(\ln f)/dp\right\} + \rho\, A^\lambda_{\cdot\alpha} U^\alpha = 0, \tag{1.9}$$

and

$$S^\alpha{}_{,\alpha} = 0, \tag{1.10}$$

as the governing equations. If we were to set $a = 0$ and $b = 1$, we would obtain the equations of VI-5, which describe Maxwell–Lorentz electrodynamics; and if we were to set $b = \rho = 0$ we would obtain the equations of VI-6, which describe the Einstein gravitation theory in the absence of electromagnetic effects. Since the above system of equations results from a linear combination of the Lagrangian functions representing these two basic fields, it is natural to make the tentative assumption that a description of interacting electromagnetic and gravitational fields in the presence of matter is provided by equations (1.6)–(1.9). Under this assumption, (1.6) describes the electromagnetic field, (1.8) describes the gravitational field, and (1.9) describes the fibers of macroscopic matter.

In order to ensure that we consider only $I^{\alpha\beta}$ fields of the first case, that is, $I^{\alpha\beta}$ fields of the nonnull class, we make the following assumption.

CONSTITUTIVE ASSUMPTION 1.1. *It is explicitly assumed that* a, b, *and* $I^{\alpha\beta}{}_{;\beta}$ *are nonzero.*

Under this constitutive assumption, the hypotheses of Theorem VI-4.2 are satisfied. The field space $\mathfrak{F}$ is thus well based, and equations (1.7) are equivalent to

$$h_{\alpha\beta;\gamma} = \tfrac{1}{2}\, h_{\alpha\beta}\, f_\gamma - h_{\gamma(\alpha}\, f_{\beta)}. \tag{1.11}$$

From this it may easily be verified that

$$(\sqrt{-h}\; h^{\alpha\beta})_{;\beta} = \tfrac{5}{2}\sqrt{-h}\, f^{\alpha}_{\cdot}. \tag{1.12}$$

Setting $dp/ds = 1$, that is, setting $f^2 = 1$, which was shown in VI-3 to be possible with no loss of generality, and applying Theorem VI-4.2, we have the following result.

THEOREM 1.1. *If* $\mathfrak{F}$ *is a field space for which Constitutive Assumptions* VI-2.1, VI-2.2, VI-4.1, *and* 1.1 *are satisfied, then the affine connection is given by*

$$L^{\alpha}{}_{\beta\gamma} = \Gamma^{\alpha}{}_{\beta\gamma}(h_{\eta\pi}) - \tfrac{1}{4}\,(f_\gamma\, \delta^{\alpha}_{\beta} + f_\beta\, \delta^{\alpha}_{\gamma} - 3 f^{\alpha}_{\cdot}\, h_{\beta\gamma}) \tag{1.13}$$

in terms of the 14 potential-like functions $h_{\alpha\beta}$ *and* f_γ, *where these functions must satisfy the system of field equations*

$$b\,(\sqrt{-h}\; A^{\alpha\beta}_{\cdot\cdot})_{,\beta} = -\frac{3a}{2}\sqrt{-h}\, f^{\alpha}_{\cdot} + \rho\, U^{\alpha}, \tag{1.14}$$

$$A_{\gamma\lambda} \overset{\text{def}}{=} f_{[\gamma,\lambda]}, \tag{1.15}$$

$$\mu\, U^{\beta}\, \nabla_{\beta} U^{\lambda} + \rho\, A^{\lambda}_{\cdot\alpha}\, U^{\alpha} = 0, \tag{1.16}$$

$$a\left(R^{\alpha\beta}_{\cdot\cdot} - \frac{1}{2}R\,h^{\alpha\beta}\right) - \frac{\Lambda}{2}h^{\alpha\beta} = -\frac{\mu}{2}(-h)^{-1/2}\,U^{\alpha}\,U^{\beta}$$

$$+\frac{b}{2}\left(A^{\alpha\eta}_{\cdot\cdot}\,A^{\beta}_{\cdot\eta} - \frac{1}{4}h^{\alpha\beta}\,A_{\lambda\eta}\,A^{\lambda\eta}_{\cdot\cdot}\right), \qquad (R = R_{\alpha\beta}\,h^{\alpha\beta}), \tag{1.17}$$

$$R_{\eta\lambda} = J_{\eta\lambda} - \tfrac{3}{4}\,h_{\eta\lambda}\,\nabla_{\rho}f^{\rho}_{\cdot} + \tfrac{3}{8}\,f_{\eta}\,f_{\lambda}. \tag{1.18}$$

It is of interest to examine the degree of determinism of the system of field equations obtained in the above Theorem 1.1. The quantities to be determined are the four f_{α}, the four U^{α}, the ten $h_{\alpha\beta}$, and the two functions μ and ρ, giving a total of twenty. We have four equations from the system (1.14) and (1.15), four from the system (1.16), and ten from the system (1.17) and (1.18), giving a total of eighteen. The system would thus seem to be indeterministic. It will be shown, however, that the differential identities of the curvature tensor $J_{\alpha\beta}$ and the equation $S^{\alpha}{}_{,\alpha} = 0$ imply that any solution of the field equations (1.14)–(1.18) is such that all twenty functions are determined.

2. Scaling and Normalization

The field equations (1.14)–(1.18) established in Theorem 1.1 relate a system of field variables over the abstract field space $\mathfrak{F}$ under the satisfaction of Constitutive Assumptions VI-2.1, VI-2.2, VI-4.1, and 1.1. In order to state these equations in a more familiar and recognizable form, it is useful to introduce certain scalings and normalizations of the field variables in this abstract space.

We first rewrite the equations (1.14)–(1.18) in the following obvious form:

$$(\sqrt{-h}\,A^{\alpha\beta}_{\cdot\cdot})_{,\beta} = -\frac{3a}{2b}\sqrt{-h}\,f^{\alpha}_{\cdot} + \frac{\rho}{b}\,U^{\alpha}, \tag{2.1}$$

$$A_{\alpha\beta} = f_{[\alpha,\beta]}, \tag{2.2}$$

$$\mu\,U^{\beta}\,\nabla_{\beta}U^{\lambda} + \rho\,A^{\lambda}_{\cdot\alpha}\,U^{\alpha} = 0, \tag{2.3}$$

$$R_{\alpha\beta} - \frac{1}{2}R\,h_{\alpha\beta} - \frac{\Lambda}{2a}h_{\alpha\beta} = -\frac{\mu}{2a\sqrt{-h}}\,U_{\dot\alpha}\,U_{\dot\beta}$$

$$+\frac{b}{2a}\left(A_{\alpha\cdot}^{\ \eta}\,A_{\beta\eta} - \frac{1}{4}h_{\alpha\beta}\,A_{\lambda\eta}\,A^{\lambda\eta}_{\cdot\cdot}\right), \tag{2.4}$$

where the (α, β) indices have been lowered in (1.17) by use of the tensor $h_{\eta\pi}$. We have seen in IV-4 that the affine connection $L^{\alpha}{}_{\beta\gamma}$ is invariant under a general class of simultaneous conformal, projective, and gauge transformations; in particular, it is invariant under the conformal transformation (normalization)

$$'h_{\alpha\beta} = \sigma\, h_{\alpha\beta}, \tag{2.5}$$

where σ = constant. (See IV-5, where the conformal transformations on the $I^{\alpha\beta}$ field are shown to be equivalent to (2.5) when the field Lagrangian function is a scalar density.) Hence $R_{\alpha\beta}$ and $A_{\alpha\beta}$ are invariant under (2.5). The connection $\Gamma^{\alpha}{}_{\beta\gamma}(h_{\eta\pi})$ is also invariant under (2.5), since as a function of $h_{\alpha\beta}$ and $h_{\alpha\beta,\gamma}$ together it is homogeneous of degree zero, and since σ = constant. Substituting (2.5) into (2.1)–(2.4), setting

$$'f^2 = 'h_{\alpha\beta}\, U^{\alpha}\, U^{\beta},$$

and *dropping the primes* gives

$$(\sqrt{-h}\, A^{\alpha\beta}_{\cdot\cdot})_{,\beta} = -\frac{3a}{2b\,\sigma}\sqrt{-h}\, f^{\alpha}_{\cdot} + \frac{\rho}{b}\, U^{\alpha}, \tag{2.6}$$

$$A_{\alpha\beta} = f_{[\alpha,\beta]}, \tag{2.7}$$

$$\mu\, U^{\beta}\, \nabla_{\beta} U^{\lambda} + \rho\, \sigma\, A^{\gamma}_{\cdot\alpha}\, U^{\alpha} = 0, \tag{2.8}$$

$$R_{\alpha\beta} - \frac{1}{2} R\, h_{\alpha\beta} - \frac{\Lambda}{2\sigma\, a} h_{\alpha\beta} = -\frac{\mu}{2a\sqrt{-h}} U_{\dot\alpha}\, U_{\dot\beta} + \frac{b\,\sigma}{2a}\left(A_{\alpha\eta}\, A_{\beta\cdot}^{\cdot\eta} - \frac{1}{4} h_{\alpha\beta}\, A_{\lambda\eta}\, A^{\lambda\eta}_{\cdot\cdot}\right), \tag{2.9}$$

$$f^2 = h_{\alpha\beta}\, U^{\alpha}\, U^{\beta} = \sigma. \tag{2.10}$$

Since f^2 is no longer unity, we have

$$\frac{ds}{dp} = f = \sigma^{1/2}.$$

Thus, setting

$$V^{\alpha} = \frac{dx^{\alpha}}{ds}, \tag{2.11}$$

we obtain

$$U^\alpha = \frac{dx^\alpha}{dp} = \frac{dx^\alpha}{ds}\frac{ds}{dp} = f\,V^\alpha = \sigma^{1/2}\,V^\alpha. \tag{2.12}$$

In order to interpret the resulting equations, we shall find it necessary to take s rather than p as the independent variable, since it is s that is used in the theories of Einstein and others. When we set

$$\frac{ds}{dp} = f = \sigma^{1/2},$$

the above equations reduce to

$$(\sqrt{-h}\,A^{\alpha\beta}_{\cdot\cdot})_{,\beta} = -\frac{3a}{2b\,\sigma}\sqrt{-h}\,f^\alpha_\cdot + \frac{\rho\sqrt{\sigma}}{b}\,V^\alpha, \tag{2.13}$$

$$A_{\alpha\beta} = f_{[\alpha,\beta]}, \tag{2.14}$$

$$\mu\,V^\beta\,\nabla_\beta V^\lambda + \sqrt{\sigma}\,\rho\,h^{\lambda\eta}\,A_{\eta\alpha}\,V^\alpha = 0, \tag{2.15}$$

$$R_{\alpha\beta} - \frac{1}{2}\,R\,h_{\alpha\beta} - \frac{\Lambda}{2\sigma\,a}\,h_{\alpha\beta} = -\frac{\mu\,\sigma}{2a\sqrt{-h}}\,V_{\dot\alpha}\,V_{\dot\beta} + \frac{b\,\sigma}{2a}\left(A_{\alpha\eta}\,A_{\beta\cdot}^{\ \eta} - \frac{1}{4}\,h_{\alpha\beta}\,A_{uv}\,A^{uv}_{\cdot\cdot}\right), \tag{2.16}$$

$$g^2 = h_{\alpha\beta}\,V^\alpha\,V^\beta = 1, \qquad V^\alpha = \frac{dx^\alpha}{ds}. \tag{2.17}$$

If we now set

$$a = b = 1, \qquad f_\alpha = \lambda\,\varphi_\alpha, \qquad A_{\alpha\beta} = \lambda\,{}^*A_{\alpha\beta}, \qquad {}^*A_{\alpha\beta} = \varphi_{[\alpha,\beta]},$$

$$\tilde\mu = X\,\mu, \qquad \tilde\Lambda = -\frac{\Lambda}{\sigma}, \qquad \tilde\rho = X\,\rho\,\lambda\sqrt{\sigma}, \tag{2.18}$$

$$\lambda^2 = \frac{2\kappa^2\,\xi}{3} = \frac{1}{X} = \frac{\xi}{\sigma},$$

and drop the tildes, after algebraic simplification we obtain

$$(\sqrt{-h}\,{}^*A^{\alpha\beta}_{\cdot\cdot})_{,\beta} = -\kappa^2\sqrt{-h}\,\varphi^\alpha_\cdot + \rho\,V^\alpha,$$

$${}^*A_{\alpha\beta} = \varphi_{[\alpha,\beta]},$$

$$\mu\,V^\beta\,\nabla_\beta V^\lambda + \rho\,h^{\lambda\eta}\,{}^*A_{\eta\alpha}\,V^\alpha = 0,$$

$$R_{\alpha\beta} - \frac{1}{2} R\, h_{\alpha\beta} + \frac{\Lambda}{2} h_{\alpha\beta} = -\frac{\xi\mu}{2\sqrt{-h}} V_{\dot\alpha} V_{\dot\beta}$$

$$+ \frac{\xi}{2}\left({}^*A_{\alpha\eta}\, {}^*A_{\beta\cdot}^{\ \eta} - \frac{1}{4} h_{\alpha\beta}\, {}^*A_{\eta\lambda}\, {}^*A^{\eta\lambda}_{\cdot\cdot}\right).$$

Substituting these results into the remaining field equations of Theorem 1.1, we have the following result.

THEOREM 2.1. *Let*

(i) $\mathfrak{F}$ *be a field space for which Constitutive Assumptions* VI-2.1, VI-2.2, VI-4.1, *and* 1.1 *are satisfied;*

(ii) *the* $h_{\alpha\beta}$ *field be normalized by the conformal transformation*

$$'h_{\alpha\beta} = \sigma\, h_{\alpha\beta}, \tag{2.19}$$

where σ *is a constant,*

$$'f^2 = 'h_{\alpha\beta}\, U^\alpha\, U^\beta,$$

and then delete all primes from the resulting equations;

(iii) p *be replaced as independent fiber parameter by the parameter* s, *which is related to* p *by*

$$\frac{ds}{dp} = \sigma^{1/2}, \tag{2.20}$$

so that

$$U^\alpha = \sigma^{1/2}\, V^\alpha, \tag{2.21}$$

where

$$V^\alpha = \frac{dx^\alpha}{ds}; \tag{2.22}$$

and

(iv) *the following scalings be made:*

$$f_\alpha = \lambda\, \varphi_\alpha, \qquad A_{\alpha\beta} = \lambda\, {}^*A_{\alpha\beta}, \qquad {}^*A_{\alpha\beta} = \varphi_{[\alpha,\beta]}, \qquad \tilde\mu = X\,\mu,$$

$$\tilde\Lambda = -\frac{\Lambda}{\sigma}, \qquad \tilde\rho = X\,\rho\,\lambda\sqrt{\sigma}, \qquad \lambda^2 = \frac{2\kappa^2\,\xi}{3} = \frac{1}{X} = \frac{\xi}{\sigma}, \tag{2.23}$$

and then delete all tildes from the resulting equations. Set $a = b = 1$, *so that* Λ *is the only arbitrary factor in the original field Lagrangian function. Then the affine connection of* $\mathfrak{F}$ *is given by*

$$L^{\gamma}{}_{\alpha\beta} = \Gamma^{\gamma}{}_{\alpha\beta}(h_{\eta\pi}) - \kappa \sqrt{\frac{\xi}{24}} \left(\varphi_\alpha \, \delta^{\gamma}_{\beta} + \varphi_\beta \, \delta^{\gamma}_{\alpha} - \varphi^{\gamma}_{\cdot} \, h_{\alpha\beta}\right) \tag{2.24}$$

in terms of the 14 potential-like functions $h_{\alpha\beta}$ and φ_α, which must satisfy the field equations

$$(\sqrt{-h} \; {}^*A^{\alpha\beta}_{\cdot\cdot})_{,\beta} = -\kappa^2 \sqrt{-h} \, \varphi^{\alpha}_{\cdot} + \rho \, V^\alpha, \tag{2.25}$$

$${}^*A_{\alpha\beta} = \varphi_{[\alpha,\beta]}, \qquad A_{\alpha\beta} = \kappa \sqrt{\frac{2\xi}{3}} \; {}^*A_{\alpha\beta}, \tag{2.26}$$

$$\mu \, V^\beta \, \nabla_\beta V^\lambda + \rho \, A^{\lambda}_{\cdot\alpha} \, V^\alpha = 0, \tag{2.27}$$

$$R_{\alpha\beta} - \frac{1}{2} R \, h_{\alpha\beta} = -\frac{\Lambda}{2} h_{\alpha\beta} - \xi \frac{\mu}{2\sqrt{-h}} V_{\dot\alpha} \, V_{\dot\beta}$$

$$+ \frac{\xi}{2} \left({}^*A_{\alpha\cdot}^{\ \ \eta} \; {}^*A_{\beta\eta} - \frac{1}{4} h_{\alpha\beta} \; {}^*A_{\lambda\eta} \; {}^*A^{\lambda\eta}_{\cdot\cdot} \right), \tag{2.28}$$

$$R_{\alpha\beta} = J_{\alpha\beta} - \sqrt{\frac{3\xi}{8}} \, \kappa \, h_{\alpha\beta} \, \nabla_\rho \varphi^{\rho}_{\cdot} + \frac{\xi \, \kappa^2}{4} \varphi_\alpha \, \varphi_\beta. \tag{2.29}$$

By this theorem, there is no loss of generality in working with the system of equations (2.25)–(2.29), rather than with the equations stated in Theorem 1.1. It is to be noted that the factors a and b have been set equal to unity, so that the resulting field Lagrangian function contains no arbitrary factors. The constants κ and ξ in (2.25)–(2.29) are thus true scaling and normalization coefficients since they are independent of a and b, and hence can be chosen arbitrarily.

3. Interpretations

We shall now establish interpretations of the various terms appearing in the scaled and normalized field equations (2.21)–(2.25) in order to determine whether they actually represent interacting electromagnetic and gravitational field phenomena. To establish such interpretations, however, it is first necessary to rewrite the field equations in such a manner that they involve only processes that are well defined in a subsidiary "physical" metric space, since it is only in such spaces that most earlier theories have been cast.

THEOREM 3.1. *If $\mathfrak{F}$ is a field space for which Constitutive Assumptions* VI-2.1, VI-2.2, VI-4.1, *and* 1.1 *are satisfied, then there is a distinct subsidiary metric space with metric tensor $h_{\alpha\beta}$ in which the field equations* (2.25)–(2.29) *assume the following form:*

$$\sqrt{-h}\,\nabla_\beta{}^*A^{\alpha\beta}_{\cdot\cdot} = -\kappa^2\sqrt{-h}\,\varphi^\alpha_\cdot + \rho\,V^\alpha, \tag{3.1}$$

$$^*A_{\alpha\beta} = \nabla_{[\beta}\varphi_{\alpha]}, \tag{3.2}$$

$$\mu\,V^\beta\,\nabla_\beta V^\lambda + \rho\,{}^*A^\lambda_{\cdot\alpha}\,V^\alpha = 0, \qquad V^\alpha = \frac{dx^\alpha}{ds}, \tag{3.3}$$

$$\begin{aligned} G_{\alpha\beta} = &-\frac{\Lambda}{2}\,h_{\alpha\beta} - \frac{\xi\,\mu}{2\sqrt{-h}}\,V_{\dot\alpha}\,V_{\dot\beta} - \sqrt{\frac{3\xi}{8}}\,\kappa\,h_{\alpha\beta}\,\nabla_\rho\varphi^\rho_\cdot \\ &+\frac{\xi}{2}\left({}^*A_{\alpha\eta}\,{}^*A_{\beta\cdot}^{\ \eta} - \frac{1}{4}\,h_{\alpha\beta}\,{}^*A_{\lambda\eta}\,{}^*A^{\lambda\eta}_{\cdot\cdot}\right) \\ &-\frac{\xi\,\kappa^2}{4}\left(\varphi_\alpha\,\varphi_\beta - \frac{1}{2}\,h_{\alpha\beta}\,\varphi_\rho\,\varphi^\rho_\cdot\right), \end{aligned} \tag{3.4}$$

where

$$G_{\alpha\beta} = J_{\alpha\beta} - \tfrac{1}{2}\,J\,h_{\alpha\beta} \tag{3.5}$$

is the Einstein tensor.

Proof. By Constitutive Assumption VI-2.1, there is defined in the abstract field space $\mathfrak{F}$ a symmetric tensor $h_{\alpha\beta}$ with nonvanishing determinant. Denote by ∇_ρ the covariant derivative formed from $\Gamma^\alpha{}_{\beta\gamma}(h_{\eta\pi})$, the Christoffel symbols of the second kind based on $h_{\alpha\beta}$. We then have

$$\nabla_\gamma h_{\alpha\beta} = 0,$$

so that the postulates M_1, M_2, M_3 of a metric space are satisfied. Since $\Gamma^\alpha{}_{\beta\gamma}(h_{\eta\pi})$ is not the affine connection of the abstract field space $\mathfrak{F}$, as seen by comparison with (1.3), this metric space is a distinct subsidiary space that is not to be identified with the field space $\mathfrak{F}$. Now $\sqrt{-h}\;{}^*A^{\alpha\beta}_{\cdot\cdot}$ is a skew-symmetric tensor density, so that

$$(\sqrt{-h}\;{}^*A^{\alpha\beta}_{\cdot\cdot})_{,\beta} \equiv \nabla_\beta(\sqrt{-h}\;{}^*A^{\alpha\beta}_{\cdot\cdot}),$$

which in turn is equal to $\sqrt{-h}\ \nabla_\beta {}^*A^{\alpha\beta}_{\cdot\cdot}$ since ∇_β annuls $h_{\alpha\beta}$ and hence $\sqrt{-h}$. Similarly, we have

$$\varphi_{[\alpha,\beta]} \equiv \nabla_{[\beta}\varphi_{\alpha]}.$$

Substituting (2.29) into (2.28) gives

$$R_{\alpha\beta} - \frac{1}{2} R\, h_{\alpha\beta} = G_{\alpha\beta} + \sqrt{\frac{3\xi}{8}}\, \kappa\, h_{\alpha\beta}\, \Delta_\rho \varphi^\rho_\cdot$$

$$+ \frac{\xi\,\kappa^2}{4}\left(\varphi_\alpha\, \varphi_\beta - \frac{1}{2}\varphi_\rho\, \varphi^\rho_\cdot\, h_{\alpha\beta}\right), \tag{3.6}$$

where $G_{\alpha\beta}$ is the Einstein tensor defined by (3.5). Substituting these results into the field equations (2.25)–(2.28) of Theorem 2.1 then yields (3.1)–(3.5), in which no processes occur except those that are well defined in the subsidiary metric space.

It is to be specifically noted that the metric space referred to in Theorem 3.1 is distinct from the field space $\mathfrak{F}$, and that ${}^*A_{\alpha\beta}$ is a well-defined curvature field in field space. In the subsidiary metric space, ${}^*A_{\alpha\beta}$ becomes a field defined over that space by (3.2) in terms of the vector function φ_α and has nothing whatsoever to do with curvature expressions in that space. Similarly, $R_{\alpha\beta}$ is a well-defined curvature field in the abstract field space, and $R_{\alpha\beta}$ is related by (2.29) to curvature expressions and functions of φ_α and its derivatives in the subsidiary metric space. The additional terms that appear on the right-hand side of (2.29) and do not depend on $J_{\alpha\beta}$ are of particular significance; they will be seen to generate just the required expressions for the representations of certain elementary phenomena.

The equations established in Theorem 3.1 are now easily interpretable. We begin by establishing the following asymptotic correspondence.

Theorem 3.2. *If $\mathfrak{F}$ is a field space for which Constitutive Assumptions* VI-2.1, VI-2.2, VI-4.1, *and* 1.1 *are satisfied, then in the limit as κ approaches zero the solution manifold of the variant field equations over the subsidiary metric space established in Theorem* 3.1 *is isomorphic with the solution manifold of the Einstein–Maxwell theory of interacting gravitational and electromagnetic fields in the presence of macroscopic matter, provided ξ and Λ are set equal to the matter-field interaction coefficient and the cosmological constant, respectively.*

Proof. Under the hypotheses, Theorem 3.1 is applicable. Taking the limit of the field equations (3.1)–(3.4) as κ approaches zero, setting ξ equal to the matter-field interaction coefficient, and setting Λ equal to the cosmological constant, we have the field equations of the Einstein–Maxwell theory:†

$$\sqrt{-h}\ \nabla_\beta\ {}^*A^{\alpha\beta}_{\cdot\cdot} = \rho\ V^\alpha, \tag{3.7}$$

$$^*A_{\alpha\beta} = \nabla_{[\beta}\varphi_{\alpha]}, \tag{3.8}$$

$$\mu\ U^\beta\ \nabla_\beta U^\lambda + \rho\ A^\lambda_{\cdot\alpha}\ V^\alpha = 0, \tag{3.9}$$

$$G_{\alpha\beta} = -\Lambda\ h_{\alpha\beta} - \frac{\mu\ \xi}{2\sqrt{-h}}\ V_{\dot\alpha}\ V_{\dot\beta} + \frac{\xi}{2}\left({}^*A_{\alpha\eta}\ {}^*A_{\beta\cdot}^{\ \eta} - \frac{1}{4}\ h_{\alpha\beta}\ {}^*A_{uv}\ {}^*A^{uv}_{\cdot\cdot}\right). \tag{3.10}$$

This establishes the theorem.

Theorems 2.1 and 3.1 result from the normalization

$$'h_{\alpha\beta} = \sigma\ h_{\alpha\beta}.$$

Further, by (2.23), σ approaches infinity as κ approaches zero. Hence, remembering that we have dropped the primes so that the $h_{\alpha\beta}$ appearing after Theorem 2.1 are in actuality $'h_{\alpha\beta}$, we see that all the components of the original $h_{\alpha\beta}$ field must be proportional to κ if the metric tensor in Theorem 3.1 is to be asymptotically finite. Since the original $h_{\alpha\beta}$ field was an arbitrary symmetric tensor defined over the abstract field space $\mathfrak{F}$, it can with no loss of generality be chosen proportional to κ. This choice, however, is equivalent to requiring the original $h_{\alpha\beta}$ in field space to approach zero as κ approaches zero, and hence *the above theorem represents a case in which there is significant decoupling between the $A_{\alpha\beta}$ field and the $R_{\alpha\beta}$ field.* (From inspection of the form of the Lagrangian function (1.1), the term containing $R_{\alpha\beta}$ involves $h_{\alpha\beta}$ to the first power, compared with the term containing $A_{\alpha\beta}$, which involves $h_{\alpha\beta}$ to the second power, to within the common factor $\sqrt{-h}$. This situation is reminiscent of what occurs in the theory of "bare" particles and renormalization as it is currently inter-

† See R. C. Tolman: *Relativity Thermodynamics and Cosmology*, Clarendon Press, Oxford (1934), pp. 259–290, for a relatively complete treatment of the Einstein–Maxwell theory.

preted. It has one advantage, however, in that our original $h_{\alpha\beta}$ field was arbitrary and not tied to any pre-envisioned physical interpretation.) This is also evident if one examines (2.24); namely, the affine connection of the abstract field space goes asymptotically to the affine connection of the metric space as κ approaches zero. Combining (2.24) and $(2.26)_2$, however, we have

$$*A_{\alpha\beta} = \frac{1}{\kappa}\sqrt{\frac{3}{2\xi}}\,A_{\alpha\beta} = \frac{1}{\kappa}\sqrt{\frac{3}{2\xi}}\,L^{\sigma}{}_{\sigma[\beta,\alpha]} = \varphi_{[\alpha,\beta]},$$

so that the $*A_{\alpha\beta}$ field does not vanish with κ.

Theorem 3.2 establishes the results anticipated at the beginning of this chapter, namely, that *interacting gravitational and electromagnetic fields in the presence of macroscopic matter are represented by the variant field equations under Constitutive Assumptions* VI-2.1, VI-2.2, VI-4.1, *and* 1.1. *Having established this result by use of the subsidiary metric space of Theorem* 3.1, *we may then revert to the abstract field space as governed by the system of equations obtained in Theorem* 2.1, *where both the electromagnetic and gravitational fields arise from curvature expressions, and hence from a common geometric basis.*

This is by no means the first instance in which the Einstein–Maxwell or just the Einstein theory results from a limiting case in which the quantity κ goes to zero. Several recent papers concerned with the quantum mechanical aspects of field theory have shown that there are cases in which the Einstein–Maxwell or the Einstein theory results when the field mass approaches zero.†

The following theorem establishes the basis for interpreting the general case.

THEOREM 3.3. *If* $\mathfrak{F}$ *is a field space for which Constitutive Assumptions* VI-2.1, VI-2.2, VI-4.1, *and* 1.1 *are satisfied, and if* $\rho = \mu = 0$, *then the solution manifold of the variant field equations over the subsidiary metric space established in Theorem* 3.1 *is isomorphic with the solution manifold of the Einstein gravitational field interacting with the (unquantized) vector-meson field with field mass equal to* κ, *in which* ξ *and* Λ *are set equal to the matter-field interaction coefficient and the cosmological constant, respectively.* (The term "vector-meson field" is used here in a generic sense. It is not intended to imply that the field mass κ is necessarily the field mass occurring in the quantum field theory, but

† O. Brulin and S. Hjalmars: "The Gravitational Zero Mass Limit of Spin-2 Particles," *Arkiv för Fysik*, **16** (1959), pp. 19–32; "An Alternative Formulation of the Linearized Classic Theory of Gravitation as a Zero Mass Limit," *Arkiv fèr Fysik*, **18** (1960), pp. 209–217.

only that the resulting equations are of the same nature and structure as those encountered in the study of the unquantized vector-meson field.)

Proof. Under the hypotheses, Theorem 3.1 is applicable. Setting $\rho = \mu = 0$ in (3.1)–(3.4), we have

$$\sqrt{-h}\,\nabla_\beta\ {}^*A^{\alpha\beta}_{\cdot\cdot} = -\sqrt{-h}\,\kappa^2\,\varphi^{\alpha}_{\cdot}, \tag{3.11}$$

$$ {}^*A_{\alpha\beta} = \nabla_{[\beta}\varphi_{\alpha]}, \tag{3.12}$$

and

$$\begin{aligned} G_{\alpha\beta} = {} & -\Lambda\, h_{\alpha\beta} - \sqrt{\frac{3\xi}{8}}\,\kappa\, h_{\alpha\beta}\,\nabla_\rho\varphi^{\rho}_{\cdot} \\ & + \frac{\xi}{2}\left({}^*A_{\alpha\eta}\ {}^*A_{\beta\cdot}^{\ \eta} - \frac{1}{4}\,{}^*A_{\lambda\eta}\ {}^*A^{\lambda\eta}_{\cdot\cdot}\,h_{\alpha\beta}\right) \\ & - \frac{\kappa^2\,\xi}{4}\left(\varphi_\alpha\,\varphi_\beta - \frac{1}{2}\,h_{\alpha\beta}\,\varphi_\rho\,\varphi^{\rho}_{\cdot}\right). \end{aligned} \tag{3.13}$$

Since (3.11) can be written as

$$(\sqrt{-h}\ {}^*A^{\alpha\beta}_{\cdot\cdot})_{,\beta} = -\sqrt{-h}\,\kappa^2\,\varphi^{\alpha}_{\cdot},$$

we have

$$(\sqrt{-h}\,\kappa^2\,\varphi^{\alpha}_{\cdot})_{,\alpha} = 0$$

as a consequence of the skew-symmetry of ${}^*A^{\alpha\beta}_{\cdot\cdot}$; but $\sqrt{-h}\,\varphi^{\alpha}_{\cdot}$ is a vector density, so that

$$(\sqrt{-h}\,\varphi^{\alpha}_{\cdot})_{,\alpha} = \sqrt{-h}\,\nabla_\alpha\varphi^{\alpha}_{\cdot},$$

and hence

$$\nabla_\alpha\varphi^{\alpha}_{\cdot} = 0. \tag{3.14}$$

Substituting (3.14) into (3.13) then gives

$$\begin{aligned} G_{\alpha\beta} = {} & -\Lambda\, h_{\alpha\beta} - \frac{\kappa^2\,\xi}{4}\left(\varphi_\alpha\,\varphi_\beta - \frac{1}{2}\,h_{\alpha\beta}\,\varphi_\rho\,\varphi^{\rho}_{\cdot}\right) \\ & + \frac{\xi}{2}\left({}^*A_{\alpha\eta}\ {}^*A_{\beta\cdot}^{\ \eta} - \frac{1}{4}\,h_{\alpha\beta}\,{}^*A_{\lambda\eta}\ {}^*A^{\lambda\eta}_{\cdot\cdot}\right). \end{aligned} \tag{3.15}$$

Equations (3.11), (3.12), and (3.15) describe the vector-meson field with field mass κ interacting with the Einstein gravitational field.†

The origin of the term

$$\frac{\kappa^2 \xi}{4}\left(\varphi_\alpha \varphi_\beta - \frac{1}{2} h_{\alpha\beta} \varphi_\rho \varphi^\rho_\cdot\right)$$

in (3.15), which is the kinetic energy of the vector-meson field, is of particular interest.

THEOREM 3.4. *If* $\mathfrak{F}$ *is a field space for which Constitutive Assumptions* VI-2.1, VI-2.2, VI-4.1, *and* 1.1 *are satisfied, then in the subsidiary metric space established in Theorem* 3.1 *the kinetic energy of the vector-meson field is a direct consequence of the* $R_{\alpha\beta}$ *field and is always present except in the asymptotic case in which* κ *goes to zero.*

Proof. By (3.6), we have

$$R_{\alpha\beta} - \frac{1}{2} R\, h_{\alpha\beta} = G_{\alpha\beta} + \sqrt{\frac{3\xi}{8}}\, \kappa\, h_{\alpha\beta}\, \nabla_\rho \varphi^\rho_\cdot$$

$$+ \frac{\xi\, \kappa^2}{4}\left(\varphi_\alpha \varphi_\beta - \frac{1}{2} \varphi_\rho \varphi^\rho_\cdot h_{\alpha\beta}\right).$$

Thus, unless κ goes to zero, the term

$$\frac{\xi\, \kappa^2}{4}\left(\varphi_\alpha \varphi_\beta - \frac{1}{2} \varphi_\rho \varphi^\rho_\cdot h_{\alpha\beta}\right)$$

is always present.

Theorem 3.4 establishes a result in significant contrast to the usual procedure used in obtaining the vector-meson field. What is generally done is to include a term of the type $\kappa^2 \sqrt{-h}\, \varphi_\alpha \varphi^\alpha_\cdot$ directly in the Lagrangian function for the field.‡ Including this, however, is equivalent to knowing what the result should be and rigging the mathematics accordingly. In the case of the variant field theory, the Lagrangian function contains no such term. The kinetic-energy expression results directly from

† B. Hoffman: "The Vector Meson Field and Projective Relativity," *Phys. Rev.*, **72** (1947), pp. 458–465.

‡ S. S. Schweber, H. A. Bethe, and F. de Hoffman: *Mesons and Fields. Vol. I, Fields*, Row, Peterson and Company, White Plains, New York (1955), p. 107, eq. 76.

the curvature expression $R_{\alpha\beta}$ in the abstract field space when the expression for $R_{\alpha\beta}$ is rewritten for the subsidiary metric space, and similarly the term on the right-hand side of equation (3.11) results directly from the curvature expression $A_{\alpha\beta}$ in the abstract field space when the Lagrangian function is varied in accordance with axiom F_4. In this sense, the vector-meson field structure is inherent in the variant field theory with the Lagrangian function (1.1) and is a derived, rather than an assumed, property. It has its origin in the structure of the abstract field space and arises whenever directly interacting $R_{\alpha\beta}$ and $A_{\alpha\beta}$ fields are considered in conjunction with the previous constitutive assumption.

It is of interest to note that the three constants Λ, κ, and ξ are independent and can be assigned at will. This is in significant contrast to a majority of recent theories in which vector-meson structure has resulted from geometrical considerations, in that there are usually less than three constants involved. In most cases there is only one, with the attendant problem of how to resolve the magnitude of the meson mass with the magnitude of the cosmological constant and the matter-field interaction coefficient.†

Combining the above results, we state the following:

If $\mathfrak{F}$ *is a field space for which Constitutive Assumptions* VI-2.1, VI-2.2, VI-4.1, *and* 1.1 *are satisfied, then the solution manifold of the variant field equations over the subsidiary metric space established in Theorem* 3.1 *is interpreted as the solution manifold of the Einstein field interacting with macroscopic matter and the (unquantized) vector-meson field with field mass* κ, *in which* ξ *and* Λ *are set equal to the matter-field interaction coefficient and the cosmological constant, respectively.*

4. On the Conservation of Mass and Charge and the Resulting Picture of Matter

It was stated in Section 1 that we have 18 field equations, plus the relations implied by the equation

$$S^{\alpha}{}_{,\alpha} = 0$$

and the differential identities satisfied by the $J_{\alpha\beta}$ field, for the determination of the 20 variables f_α, V^α, $h_{\alpha\beta}$, ρ, and μ. We now undertake the study of the differential identities of the $J_{\alpha\beta}$ field and the equation

† B. Hoffman: *op. cit.;* also "The Gravitational, Electromagnetic and Vector Meson Fields and the Similarity Geometry," *Phys. Rev.*, **73** (1948), pp. 30–35.

$$S^{\alpha}{}_{,\alpha} = 0,$$

in order to show that the solution manifold of the field equations actually determines all 20 variables. To facilitate matters, we shall assume throughout this section that the hypotheses of Theorem 3.1 are satisfied and hence equations (3.1)–(3.5) hold.

The first equation to be considered is

$$S^{\alpha}{}_{,\alpha} = 0.$$

Under the scaling and normalization established in Theorem 2.1, the Maxwellian current S^{α} is correspondingly scaled. Since the resulting equations (2.25) read

$$(\sqrt{-h}\ {}^{*}A^{\alpha\beta}_{\cdot\cdot})_{,\beta} = -\kappa^{2}\sqrt{-h}\ \varphi^{\alpha}_{\cdot} + \rho\, V^{\alpha}, \tag{4.1}$$

we may take as the scaled Maxwellian current $'S^{\alpha}$ the expression

$$'S^{\alpha} = -\kappa^{2}\sqrt{-h}\ \varphi^{\alpha}_{\cdot} + \rho\, V^{\alpha}, \tag{4.2}$$

since we have

$$'S^{\alpha}{}_{,\alpha} = 0 \tag{4.3}$$

as a result of (4.1) and the skew-symmetry of the ${}^{*}A_{\alpha\beta}$ field. We shall refer to $'S^{\alpha}$ as the *total current.*

The occurrence of the two terms in (4.2) is in many respects similar to the classical partition of the total current into a conduction current and a convective current. The term ρV^{α} is exactly of the form that represents the macroscopic convective current in Maxwell–Lorentz electrodynamics, and, by (3.3), ρ is the macroscopic charge density, since it gives the observable Lorentz force. It is thus natural to inquire whether the term $-\kappa^{2}\sqrt{-h}\ \varphi^{\alpha}_{\cdot}$ can be interpreted as a conduction-like current. Recalling that this term arose from the term

$$\tfrac{3}{5}\,(a\sqrt{-h}\ h^{\alpha\beta})_{;\beta} = \tfrac{3}{5}\, I^{\alpha\beta}{}_{;\beta}$$

(see equation (1.6)) under the scaling and normalization established in Theorem 2.1, and that the $h_{\alpha\beta}$ are the potential functions of the gravitational field, we can, in one sense, consider $-\kappa^{2}\sqrt{-h}\ \varphi^{\alpha}_{\cdot}$ as a conduction-like current. *The conductive medium is not a physical medium, however, but rather is the gravitational field* whose potential tensor $h_{\alpha\beta}$, when combined in the form $\sqrt{-h}\, h^{\alpha\beta}$, yields the polarization density† of this conductive current.

By (4.2), $'S^\alpha$ is a vector density, so that we have

$$'S^\alpha{}_{,\alpha} \equiv \nabla_\alpha\, 'S^\alpha,$$

and hence (4.3) is equivalent to

$$\nabla_\alpha\, 'S^\alpha = 0. \tag{4.4}$$

Substituting (4.2) into (4.4), we have the following result.

THEOREM 4.1. *If Theorem 3.1 holds, then the variant field equations give the following equation for the macroscopic charge density* ρ:

$$\nabla_\alpha(\rho\, V^\alpha) = \kappa^2 \sqrt{-h}\, \nabla_\alpha \varphi^\alpha_{.} = (\rho\, V^\alpha)_{,\alpha}. \tag{4.5}$$

We now consider the differential identities satisfied by $J_{\alpha\beta}$. By Theorem 3.1, on raising the (β) index through use of $h^{\beta\gamma}$ and then setting $\gamma = \beta$, we have

$$G^\beta{}_\alpha = \frac{-\Lambda}{2}\,\delta^\beta_\alpha - \frac{\mu\,\xi}{2\sqrt{-h}}\, V_\alpha\, V^\beta - \sqrt{\frac{\xi 3}{8}}\,\kappa\,\delta^\beta_\alpha\, \nabla_\rho \varphi^\rho_{.}$$

$$+ \frac{\xi}{2}\, E^\beta{}_\alpha - \frac{\xi\,\kappa^2}{4}\left(\varphi_\alpha\, \varphi^\beta - \frac{1}{2}\,\delta^\beta_\alpha\, \varphi_\rho\, \varphi^\rho_{.}\right), \tag{4.6}$$

where

$$E^\beta{}_\alpha = {}^*A_{\alpha\eta}\, {}^*A^{\beta\eta}_{..} - \tfrac{1}{4}\,\delta^\beta_\alpha\, {}^*A_{\lambda\eta}\, {}^*A^{\lambda\eta}_{..}. \tag{4.7}$$

It was shown in II-3 that the only differential identities that the $J_{\alpha\beta}$ satisfy are

$$\nabla_\beta G^\beta{}_\alpha = 0. \tag{4.8}$$

In order to simplify the ensuing calculations we first establish the following result.

LEMMA 4.1. *We have*

$$\nabla_\beta E^\beta{}_\alpha = -\, {}^*A_{\alpha\eta}\, \nabla_\beta {}^*A^{\eta\beta}. \tag{4.9}$$

† C. Truesdell and R. Toupin: "The Classical Field Theories," *Handbuch der Physik*, Band III/1, Springer-Verlag, Berlin (1960), pp. 683–689.

Proof. As a result of the skew-symmetry of $*A^{\beta\eta}$, we have

$$*A^{\beta\eta}_{\cdot\cdot}\, \nabla_{(\beta}{*A}_{|\alpha|\eta)} \equiv 0.$$

Hence, from (4.7), we obtain

$$\begin{aligned} \nabla_\beta E^\beta{}_\alpha &= *A_{\alpha\eta}\, \nabla_\beta {*A}^{\beta\eta}_{\cdot\cdot} + *A^{\beta\eta}_{\cdot\cdot}\, \nabla_\beta {*A}_{\alpha\eta} - \tfrac{1}{2}\, *A^{\lambda\eta}\, \nabla_\alpha {*A}_{\lambda\eta} \\ &= *A_{\alpha\eta}\, \nabla_\beta {*A}^{\beta\eta}_{\cdot\cdot} + *A^{\beta\eta}_{\cdot\cdot}\, \nabla_{[\beta} {*A}_{|\alpha|\eta]} - \tfrac{1}{2}\, *A^{\lambda\eta}\, \nabla_\alpha {*A}_{\lambda\eta}. \end{aligned}$$

Thus we have

$$\nabla_\beta E^\beta{}_\alpha = *A_{\alpha\eta}\, \nabla_\beta {*A}^{\beta\eta}_{\cdot\cdot} - *A^{\beta\eta}_{\cdot\cdot}\, \nabla_{[\beta} {*A}_{\eta]\alpha} - \tfrac{1}{2}\, *A^{\lambda\eta}\, \nabla_\alpha {*A}_{\lambda\eta},$$

since

$$*A_{\alpha\eta} = -*A_{\eta\alpha},$$

and hence we obtain

$$\begin{aligned} \nabla_\beta E^\beta{}_\alpha &= *A_{\alpha\eta}\, \nabla_\beta {*A}^{\beta\eta}_{\cdot\cdot} - \tfrac{1}{2}\, *A^{\beta\eta}\, (\nabla_\beta {*A}_{\eta\alpha} - \nabla_\eta {*A}_{\beta\alpha} + \nabla_\alpha {*A}_{\beta\eta}) \\ &= *A_{\alpha\eta}\, \nabla_\beta {*A}^{\beta\eta}_{\cdot\cdot} - \tfrac{1}{2}\, *A^{\beta\eta}\, (\nabla_\beta {*A}_{\eta\alpha} + \nabla_\eta {*A}_{\alpha\beta} + \nabla_\alpha {*A}_{\beta\eta}) \\ &= *A_{\alpha\eta}\, \nabla_\beta {*A}^{\beta\eta}_{\cdot\cdot} - \frac{3!}{2}\, *A^{\beta\eta}\, \nabla_{[\beta} {*A}_{\eta\alpha]} \\ &= *A_{\alpha\eta}\, \nabla_\beta {*A}^{\beta\eta}_{\cdot\cdot} - \frac{3!}{2}\, *A^{\beta\eta}\, {*A}_{[\eta\alpha,\beta]}, \end{aligned}$$

on observing that

$$\nabla_{[\beta} {*A}_{\eta\alpha]} \equiv *A_{[\eta\alpha,\beta]}.$$

Thus, since

$$*A_{[\eta\alpha,\beta]} = 0$$

and

$$\nabla_\beta {*A}^{\beta\eta} = -\nabla_\beta {*A}^{\eta\beta}_{\cdot\cdot}$$

as a result of the skew-symmetry of $*A^{\eta\beta}$, the lemma is established.

We can now obtain the equation that governs μ. Substituting (4.6) into (4.8), we have

$$0 = -\frac{\xi}{2}\left\{\frac{\mu}{\sqrt{-h}} V^\beta \nabla_\beta V^{\cdot}_{\alpha} + \frac{V^{\cdot}_{\alpha}}{\sqrt{-h}} \nabla_\beta(\mu V^\beta)\right\}$$

$$- \sqrt{\frac{3\xi}{8}}\, \kappa\, \nabla_\alpha \nabla_\rho \varphi^{\rho}_{\cdot} + \frac{1}{2} \nabla_\beta E^{\beta}{}_{\alpha}$$

$$- \frac{\kappa^2 \xi}{4} (\varphi_\alpha \nabla_\beta \varphi^{\beta}_{\cdot} + \varphi^{\beta}_{\cdot} \nabla_\beta \varphi_\alpha - \varphi^{\rho}_{\cdot} \nabla_\alpha \varphi_\rho).$$

Using Lemma 4.1 and simplifying gives

$$0 = -\left\{\frac{\mu}{\sqrt{-h}} V^\beta \nabla_\beta V^{\cdot}_{\alpha} + \frac{V^{\cdot}_{\alpha}}{\sqrt{-h}} \nabla_\beta(\mu V^\beta)\right\}$$

$$- \sqrt{\frac{3}{2\xi}}\, \kappa\, \nabla_\alpha \nabla_\rho \varphi^{\rho}_{\cdot} - {}^*A_{\alpha\eta} \nabla_\beta {}^*A^{\eta\beta}$$

$$- \frac{\kappa^2}{2} (\varphi_\alpha \nabla_\beta \varphi^{\beta}_{\cdot} + 2\varphi^\beta \nabla_{[\beta}\varphi_{\alpha]}).$$

By (3.1) and (3.2) we have

$$\nabla_\beta {}^*A^{\eta\beta}_{\cdot\cdot} = -\kappa^2 \varphi^{\eta}_{\cdot} + \frac{\rho}{\sqrt{-h}} V^\eta$$

and

$${}^*A_{\alpha\beta} = \nabla_{[\beta}\varphi_{\alpha]}.$$

Substituting these expressions into the above equation gives

$$0 = -\frac{\mu}{\sqrt{-h}} V^\beta \nabla_\beta V^{\cdot}_{\alpha} + \frac{V^{\cdot}_{\alpha}}{\sqrt{-h}} \nabla_\beta(\mu V^\beta)$$

$$- \sqrt{\frac{3}{2\xi}}\, \kappa\, \nabla_\alpha \nabla_\rho \varphi^{\rho}_{\cdot} - {}^*A_{\alpha\eta}\left(-\kappa^2 \varphi^{\eta}_{\cdot} + \frac{\rho}{\sqrt{-h}} V^\eta\right)$$

$$- \frac{\kappa^2}{2} (\varphi_\alpha \nabla_\beta \varphi^{\beta}_{\cdot} + 2\varphi^{\beta}_{\cdot}\, {}^*A_{\alpha\beta})$$

$$= -\frac{1}{\sqrt{-h}} (\mu V^\beta \nabla_\beta V^{\cdot}_{\alpha} + \rho\, {}^*A_{\alpha\eta} V^\eta) - \frac{V^{\cdot}_{\alpha}}{\sqrt{-h}} \nabla_\beta(\mu V^\beta)$$

$$- \kappa\left(\sqrt{\frac{3}{2\xi}} \nabla_\alpha + \frac{\kappa}{2} \varphi_\alpha\right) \nabla_\rho \varphi^{\rho}_{\cdot}.$$

By (3.3), however, the first parenthesis vanishes, and hence we are left with

$$0 = V_{\cdot\alpha}\, \nabla_\beta(\mu\, V^\beta) + \frac{\kappa\sqrt{-h}}{2}\left(\kappa\, \varphi_\alpha + \sqrt{\frac{6}{\xi}}\, \nabla_\alpha\right) \nabla_\rho \varphi^{\rho}_{\cdot}.$$

Multiplying by V^α and remembering that

$$V^\alpha\, V_{\cdot\alpha} = h_{\alpha\beta}\, V^\alpha\, V^\beta = 1$$

for any V^α satisfying the matter fiber equations, we obtain

$$\nabla_\beta(\mu\, V^\beta) = -\frac{\kappa\sqrt{-h}}{2}\, V^\alpha \left(\kappa\, \varphi_\alpha + \sqrt{\frac{6}{\xi}}\, \nabla_\alpha\right) \nabla_\rho \varphi^{\rho}_{\cdot}.$$

We have thus proved the following result.

THEOREM 4.2. *If Theorem* 3.1 *holds, then the variant field equations give the following equation for the macroscopic mass density* μ*:*

$$\nabla_\beta(\mu\, V^\beta) = -\frac{\kappa\sqrt{-h}}{2}\, V^\alpha \left(\kappa\, \varphi_\alpha + \sqrt{\frac{6}{\xi}}\, \nabla_\alpha\right) \nabla_\rho \varphi^{\rho}_{\cdot} = (\mu\, V^\alpha)_{,\alpha}. \quad (4.10)$$

Combining Theorems 4.1 and 4.2 and eliminating $\nabla_\rho \varphi^{\rho}_{\cdot}$, we obtain the following result.

THEOREM 4.3. *If Theorem* 3.1 *holds, then the variant field equations imply the conservation of both the macroscopic charge density* ρ *and the macroscopic mass density* μ *if and only if either the Lorentz gauge condition*

$$\nabla_\alpha \varphi^{\alpha}_{\cdot} = 0 \quad (4.11)$$

is satisfied or the field mass κ *approaches zero. If neither of these conditions is satisfied, then* μ *and* ρ *are related by the equation*

$$\nabla_\beta(\mu\, V^\beta) + \frac{1}{2\kappa}\, V^\alpha \left(\kappa\, \varphi_\alpha + \sqrt{\frac{6}{\xi}}\, \nabla_\alpha\right) \nabla_\beta(\rho\, V^\beta) = 0. \quad (4.12)$$

Theorems 4.1, 4.2, and 4.3 have very interesting implications. Examining equations (4.5) and (4.10) we see that if the Lorentz gauge condition (4.11) is not satisfied and if the field mass does not approach zero, the macroscopic charge and mass densities ρ and μ have sources given by

$$\kappa^2 \sqrt{-h}\, \nabla_\sigma \varphi^\sigma_{.} \quad \text{and} \quad -\frac{\kappa}{2} \sqrt{-h}\, V^\alpha \left(\kappa\, \varphi_\alpha + \sqrt{\frac{6}{\xi}}\, \nabla_\alpha \right) \nabla_\sigma \varphi^\sigma_{.},$$

respectively. We have seen that for the case in which the field mass approaches zero, the electromagnetic and gravitational fields are partly decoupled and the Einstein–Maxwell theory results. In classical Maxwell and Maxwell–Lorentz electrodynamics, the Lorentz gauge condition (4.11) can always be satisfied, since the electromagnetic field equations are invariant under gauge transformations of the vector potential. That this is not the case when both the field mass and ρ are nonzero is immediately evident from inspection of (4.5). (Theorem 4.1 shows that the Lorentz gauge condition is satisfied if $\rho = 0$.) Thus in the general case the Lorentz gauge condition is satisfied only by chance happening, or by adjoining (4.11) to the field equations in an ad hoc manner. Since such an ad hoc adjoining process is neither theoretically nor esthetically pleasing, we shall assume that in general the Lorentz gauge condition is not satisfied.

From the above consideration, the source terms of the macroscopic mass and charge densities are uniquely determined once $\nabla_\sigma \varphi^\sigma_{.}$ is given, that is, once the divergence of the vector-meson field is given. Thus, *the vector-meson field may be interpreted as giving rise to creation or annihilation of macroscopic mass and charge.* In this sense, variant field theory gives a classical representation of the creation-annihilation structure of space permeated by interacting gravitational and electromagnetic fields that are not partly decoupled by κ going to zero or $\rho = 0$. Since the annihilation and creation sources are respectively proportional to κ and κ^2, which are very small numbers compared with the numerical values of ρ and μ for ponderable macroscopic matter, the amount of change in the mass and charge densities would normally be macroscopically observable only on the cosmological level. It is of interest to note that such predictions of macroscopic mass and charge creation or annihilation on the cosmological level are in many respects similar to the constant-creation hypothesis of Hoyl, Bondi, and Gold.†

This is not the first prediction of the creation-annihilation process from a classical covariant theory. Even so, it has previously been held by most investigators that such predictions arose only from quantizations of the fields. Under the predictions of the variant field theory, it is seen that the

† R. A. Lytteton and H. Bondi: "On the Physical Consequences of a General Excess of Charge," *Proc. Roy. Soc. London*, **A252** (1959), pp. 313–333. The analysis given in this paper unfortunately ignores the important case in which one obtains oscillatory solutions for the charge and current creation densities which have average values comparable to those obtained by Lytteton and Bondi.

creation of matter is a consequence of direct interactions between gravitational and vector-meson fields. If either is absent, then no such predictions are forthcoming. Under the variant field theory, both the mass and charge distributions, together with their resulting fields, are responsible for creation-annihilation of macroscopic matter, and hence such processes are both electromagnetic and gravitational in nature.

We have seen that the asymptotic form of the field equations, which is obtained when the field mass goes to zero, contains no such predictions of creation-annihilation processes. Since this case has been shown to correspond to the combined Einstein–Maxwell theory, *we see that the Einstein–Maxwell theory leaves out an essential part of the picture.* In this sense, the partial decoupling that results in letting the field mass go to zero effectively eliminates the construction of matter on a rational, rather than an ad hoc, basis.

The occurrence of the vector-meson field was shown in Theorem 3.4 to be inherent in the variant field theory under (1.1), and was a derived result rather than an assumed fact. The variant field theory under (1.1) thus leads to a representation of gravitation and electromagnetism that inherently contains annihilation-creation representations of macroscopic matter, provided neither the field mass nor the macroscopic charge density ρ is set equal to zero.

It is of significance to note that the creation-annihilation prediction cannot be obtained if the macroscopic charge density ρ vanishes everywhere, since it was shown in Theorem 4.1 that $\rho = 0$ implies

$$\nabla_\beta \varphi^\beta = 0.$$

This would imply that creation-annihilation processes occur only in the presence of macroscopic current, or rather in the presence of a field generated by macroscopic current. By Theorem 4.3, the presence of macroscopic charge and annihilation-creation processes in turn imply the presence of macroscopic mass. In fact, *Theorem* 4.3 *establishes an equation of equilibrium between charge density and mass density that is maintained by the creation-annihilation processes.* With regard to this result, we can interpret the annihilation-creation process as the agent that maintains the distribution of mass and charge in the universe, in agreement with the fundamental equilibrium equation (4.12).

When we speak of creation and annihilation, what is really meant is that energy and current have changed forms. This is seen directly from the fact that $G_{\alpha\beta}$ is equal to and defines the momentum-energy tensor, and hence the latter is covariantly conserved under the differential identities satisfied by $G_{\alpha\beta}$. In fact, it was just the covariant conservation of the momentum-energy tensor that led to Theorems 4.2 and 4.3 and hence to the creation-annihilation representations. This is also borne out by equation (4.12), which may be interpreted as an equation describing the equilibrium of mass and charge density in the presence of the vector-meson field φ_α. (Equation (4.12) yields a possible explanation of the relation between the mass, charge, and charge sign, since (4.12) is not invariant under reversal of the sign of ρ.)

The creation and/or annihilation of *macroscopic* charge density may, on first examination, seem to violate the conservation of charge. This, however, is not the case. The *total* current is defined by equation (4.2) and is made up of two terms, the convective term $\rho\, V^\alpha$ involving the macroscopic charge and the conductive term $-\kappa^2 \sqrt{-h}\, \varphi^{\alpha}_{\cdot}$. The above considerations state that the total current $'S^\alpha$ is conserved (equation (4.4)). *Hence, although macroscopic charge density may be created or destroyed, the total current is conserved.* It is thus evident that when creation or annihilation of macroscopic charge occurs, what actually takes place is an interchange between the macroscopic charge density and the conductive field current, so that the forms of the charge and current change, but the total current is conserved. This situation is directly analogous to that of the creation or annihilation of macroscopic matter, which occurs in such a way that the momentum-energy tensor is covariantly conserved. It is also to be noted that for a *static* field, the macroscopic charge ρ is conserved, in agreement with laboratory measurements.

5. Field-Field Interaction

We have seen that interacting gravitational and electromagnetic fields in the presence of macroscopic matter lead to a creation-annihilation representation of elemental matter. We now turn to the pure field-field interaction and inquire whether similar conclusions are warranted.

Constitutive Assumption 5.1. *Let there be defined over the field space a symmetric tensor $h_{\alpha\beta}$ with nonvanishing determinant and let the Lagrangian function be given by*

$$\mathfrak{L} = \sqrt{-h}\,(a\, h^{\alpha\beta} R_{\alpha\beta} - {}^{b} A_{\alpha\beta} A_{\gamma\lambda}\, h^{\alpha\gamma} h^{\beta\lambda}). \tag{5.1}$$

It is of interest to note that the Lagrangian function (5.1) is identical in form to the Lagrangian function usually encountered, for vacuum gravitation and electrodynamics, with the exception that $R_{\alpha\beta}$ and $A_{\alpha\beta}$ now are curvature expressions in the abstract field space rather than the curvature-tensor contraction and the electromagnetic field tensor in a metric space.

When we examine the results leading up to and including Theorem 2.1, it is clear that under the above constitutive assumption these results hold, and we have the following result.

THEOREM 5.1. *Let $\mathfrak{F}$ be a field space for which Constitutive Assumption* 5.1 *is satisfied, let the $h_{\alpha\beta}$ field be normalized by the equation*

$$'h_{\alpha\beta} = \sigma\, h_{\alpha\beta}, \tag{5.2}$$

and then delete the primes from the resulting field equations; let the f_α be scaled by

$$f_\alpha = \frac{1}{\sigma}\psi_\alpha \tag{5.3}$$

and set

$$\sigma = b = \sqrt{\frac{3a}{2}}\frac{1}{\kappa}, \tag{5.4}$$

where κ is a constant; then the affine connection is given by

$$L^\alpha{}_{\beta\gamma} = \Gamma^\alpha{}_{\beta\gamma}(h_{\eta\pi}) - \frac{\kappa}{\sqrt{24a}}(\psi_\gamma\,\delta^\alpha_\beta + \psi_\beta\,\delta^\alpha_\gamma - 3\psi^{\alpha}_{\cdot}\,h_{\beta\gamma}) \tag{5.5}$$

in terms of the 14 potential-like functions $h_{\alpha\beta}$ and ψ_α, which must satisfy the field equations

$$(\sqrt{-h}\;{}^*A^{\alpha\beta}_{\cdot\cdot})_{,\beta} = -\kappa^2\sqrt{-h}\,\psi^\alpha_{\cdot}, \tag{5.6}$$

$${}^*A_{\alpha\beta} = \psi_{[\alpha,\beta]} = \kappa\sqrt{\frac{2}{3a}}\,A_{\alpha\beta}, \tag{5.7}$$

$$R_{\alpha\beta} - \frac{1}{2}R\,h_{\alpha\beta} = \frac{1}{2a}({}^*A_{\alpha\eta}\,{}^*A_{\beta\cdot}^{\cdot\eta} - \tfrac{1}{4}h_{\alpha\beta}\,{}^*A_{\lambda\eta}\,{}^*A^{\lambda\eta}_{\cdot\cdot}), \tag{5.8}$$

$$R_{\alpha\beta} = J_{\alpha\beta} - \sqrt{\frac{3}{8a}}\,\kappa\,h_{\alpha\beta}\,\nabla_\rho\psi^\rho_{\cdot} + \frac{\kappa^2}{4a}\varphi_\alpha\,\varphi_\beta. \tag{5.9}$$

With the field equations thus normalized and scaled, we consider the general conformal, gauge, and projective transformations derived in IV-4, which leave the $L^\alpha{}_{\beta\gamma}$ (and hence $R_{\alpha\beta}$ and $A_{\alpha\beta}$) invariant:

$$h_{\alpha\beta} = e^{-\sigma}\, 'h_{\alpha\beta}, \qquad h^{\alpha\beta} = e^{\sigma}\, 'h^{\alpha\beta}, \qquad h = e^{-4\sigma}\, 'h,$$
$$\psi_\alpha = '\psi_\alpha + \sigma_{,\alpha}, \qquad \Gamma^\alpha{}_{\beta\gamma} = '\Gamma^\alpha{}_{\beta\gamma} - \delta^{\alpha}_{(\beta}\, \sigma_{,\gamma)}. \tag{5.10}$$

(The reason we have not previously used these more general transformations is that they *do not* transform fiber equations into fiber equations, and hence in the presence of macroscopic matter the complete system of field equations is not transformed into a system of field equations.) Substituting (5.10) into (5.6)–(5.9), remembering that $R_{\alpha\beta}$ and $A_{\alpha\beta}$ are invariant under (5.10) (that is, they depend on the primed variables in the same way they depend on the unprimed variables), and dropping the primes, we have

$$(\sqrt{-h}\ {}^*A^{\alpha\beta}_{\cdot\,\cdot})_{,\beta} = -\kappa^2 e^{-\sigma} (\psi^\alpha_{\cdot} + h^{\alpha\nu} \sigma_{,\nu}) \sqrt{-h}, \tag{5.11}$$

$${}^*A_{\alpha\beta} = \psi_{[\alpha,\beta]}, \tag{5.12}$$

$$R_{\alpha\beta} - \frac{1}{2} R\, h_{\alpha\beta} = \frac{e^\sigma}{2a} E_{\alpha\beta}, \tag{5.13}$$

$$E_{\alpha\beta} = {}^*A_{\alpha\eta}\, {}^*A_{\beta\cdot}^{\ \eta} - \tfrac{1}{4} h_{\alpha\beta}\, {}^*A_{\lambda\eta}\, {}^*A^{\lambda\eta}_{\cdot\,\cdot}, \tag{5.14}$$

$$R_{\alpha\beta} = J_{\alpha\beta} - \sqrt{\frac{3}{8a}}\, \kappa\, h_{\alpha\beta} \nabla_\rho \psi^\rho_{\cdot} + \frac{\kappa^2}{4a} \psi_\alpha \psi_\beta. \tag{5.15}$$

We now apply the same reasoning as in the proof of Theorem 3.1 and obtain the following result.

THEOREM 5.2. *Let* $\mathfrak{F}$ *be a field space for which Constitutive Assumption* 5.1 *is satisfied, let* $h_{\alpha\beta}$, ψ_α, *and* $\Gamma^\alpha{}_{\beta\gamma}$ *undergo the transformations* (5.10), *and then drop the primes. There is then a distinct subsidiary metric space with metric tensor* $h_{\alpha\beta}$, *in which the field equations* (5.6)–(5.9) *assume the form*

$$\nabla_\beta\, {}^*A^{\alpha\beta}_{\cdot\,\cdot} = -\kappa^2 e^{-\sigma} (\psi^\alpha_{\cdot} + h^{\alpha\nu} \sigma_{,\nu}), \tag{5.16}$$

$${}^*A_{\alpha\beta} = \nabla_{[\beta} \psi_{\alpha]}, \tag{5.17}$$

$$G_{\alpha\beta} = \frac{e^\sigma}{2a} E_{\alpha\beta} - \frac{\kappa^2}{4a} (\psi_\alpha \psi_\beta - \tfrac{1}{2} h_{\alpha\beta} \psi_\rho \psi^\rho_{\cdot})$$
$$- \sqrt{\frac{3}{8a}}\, \kappa\, h_{\alpha\beta} \nabla_\rho \psi^\rho_{\cdot}, \tag{5.18}$$

where

$$G_{\alpha\beta} \overset{\text{def}}{=} J_{\alpha\beta} - \tfrac{1}{2} J\, h_{\alpha\beta} \tag{5.19}$$

is the Einstein tensor, and where

$$E_{\alpha\beta} = {}^*A_{\alpha\eta}\, {}^*A_{\beta\cdot}^{\ \eta} - \tfrac{1}{4} h_{\alpha\beta}\, {}^*A_{\lambda\eta}\, {}^*A^{\lambda\eta}_{\cdot\cdot} \tag{5.20}$$

is the negative of the momentum-energy tensor of the ${}^*A_{\alpha\beta}$ *field.*

Because of the skew-symmetry of ${}^*A_{\alpha\beta}$, we can again define the scaled and transformed affine current $'S^\alpha$ by

$$'S^\alpha = -\kappa^2 e^{-\sigma} (\psi^\alpha_\cdot + h^{\alpha\nu} \sigma_{,\nu}), \tag{5.21}$$

since (5.16) yields

$$\nabla_\alpha\, 'S^\alpha = 0. \tag{5.22}$$

Substituting (5.21) into (5.22), we have the following result.

THEOREM 5.3. *If the hypotheses of Theorem* 5.2 *hold, then* ψ_α *and* σ *must be such that*

$$0 = -h^{\alpha\beta} \nabla_\alpha \nabla_\beta e^{-\sigma} + \nabla_\alpha(\psi^\alpha_\cdot e^{-\sigma}) = -\Box^2 e^{-\sigma} + \nabla_\alpha(\psi^\alpha_\cdot e^{-\sigma})$$

$$= (\nabla_\alpha \psi^\alpha_\cdot + h^{\alpha\beta} \nabla_\alpha \nabla_\beta \sigma - \{\psi^\alpha_\cdot + h^{\alpha\beta} (\nabla_\beta \sigma)\} \nabla_\alpha \sigma) e^{-\sigma}. \tag{5.23}$$

It was shown in II-3 that

$$\nabla_\alpha G^\alpha{}_\beta = 0. \tag{5.24}$$

Substituting (5.18) into (5.24) gives

$$0 = \frac{e^\sigma}{2a} \nabla_\alpha E^\alpha{}_\beta + \frac{E^\alpha{}_\beta}{2a} e^\sigma \nabla_\alpha \sigma - \sqrt{\frac{3}{8a}}\, \kappa\, \nabla_\beta \nabla_\rho \psi^\rho_\cdot$$

$$- \frac{\kappa^2}{4a} (\psi_\beta \nabla_\alpha \psi^\alpha_\cdot + \psi^\alpha_\cdot \nabla_\alpha \psi_\beta - \psi^\rho_\cdot \nabla_\beta \psi_\rho).$$

Applying Lemma 4.1 and using (5.16) in the same manner as in the proof of Theorem 4.2, we have

$$0 = \frac{\kappa^2}{2a} h^{\alpha\nu}\, {}^*A_{\beta\nu} \nabla_\alpha \sigma + \frac{E^\alpha{}_\beta}{2a} e^\sigma \nabla_\alpha \sigma$$

$$- \kappa \left(\sqrt{\frac{3}{8a}} \nabla_\beta + \frac{\kappa}{2a} \psi_\beta \right) (\nabla_\alpha \psi^\alpha_\cdot).$$

The dependence on $\nabla_\alpha \psi^\alpha_{.}$ can be eliminated by using equation (5.23) of Theorem 5.3, to obtain

$$0 = \frac{h^{\alpha\nu}}{2a} (\kappa^2 \, {}^*A_{\beta\nu} + e^\sigma \, E_{\beta\nu})(\nabla_\alpha \sigma)$$

$$- \kappa \left(\sqrt{\frac{3}{8a}} \nabla_\beta + \frac{\kappa}{2a} \psi_\beta \right) (\psi^\alpha_{.} + h^{\alpha\nu} (\nabla_\nu \sigma) - h^{\alpha\nu} \nabla_\nu)(\nabla_\alpha \sigma).$$

We have thus proved the following result.

THEOREM 5.4. *If Theorem 5.2 holds, then ψ_α and σ must be such that*

$$0 = h^{\alpha\nu} (\kappa^2 \, {}^*A_{\beta\nu} + e^\sigma \, E_{\beta\nu})(\nabla_\alpha \sigma)$$

$$- \kappa \left(\sqrt{\frac{3a}{2}} \nabla_\beta + \kappa \, \psi_\beta \right) (\psi^\alpha_{.} + h^{\alpha\nu} (\nabla_\nu \sigma) - h^{\alpha\nu} \nabla_\nu)(\nabla_\alpha \sigma). \quad (5.25)$$

The following result combines Theorems 5.3 and 5.4.

THEOREM 5.5. *If the hypotheses of Theorem 5.2 are satisfied, then ψ_α and σ must be such that*

$$\nabla_\alpha \psi^\alpha_{.} = (\psi^\alpha + h^{\alpha\beta} (\nabla_\beta \sigma) - h^{\alpha\beta} \nabla_\beta)(\nabla_\alpha \sigma), \quad (5.26)$$

$$h^{\alpha\nu} (\kappa^2 \, {}^*A_{\beta\nu} + e^\sigma \, E_{\beta\nu})(\nabla_\alpha \sigma)$$

$$= \kappa \left(\sqrt{\frac{3a}{2}} \nabla_\beta + \kappa \, \psi_\beta \right) (\psi^\alpha_{.} + h^{\alpha\nu} (\nabla_\nu \sigma) - h^{\alpha\nu} \nabla_\nu)(\nabla_\alpha \sigma). \quad (5.27)$$

The basis of interpretation of the field equations (5.16)–(5.18) and the subsidiary relations (5.26) and (5.27) is the following result.

THEOREM 5.6. *A particular class of solutions to the system* (5.16), (5.17), (5.18), (5.26), *and* (5.27) *is given by $\sigma = \sigma_0 = 0$, together with the class of solutions to the field equations* (3.11)–(3.13) *of the Einstein gravitational field interacting with the (unquantized) vector-meson field with spin* = 1 *and field mass equal to κ, in which a^{-1} is set equal to the matter-field interaction coefficient.*

Proof. With $\sigma = \sigma_0 = 0$, (5.27) is identically satisfied, and (5.26) implies

$$\nabla_\alpha \psi^\alpha_{.} = 0. \quad (5.28)$$

Substituting (5.28) and $\sigma = 0$ into (5.16)–(5.18) gives equations (3.11)–(3.13) of Theorem 3.3, from which the result follows.

Since we have shown that with $\sigma = 0$ we have the classic (unquantized) vector-meson field interacting with the gravitational field, we interpret equations (5.16)–(5.18) as the equations describing (unquantized) elemental matter under the processes of creation and annihilation. For $\sigma =$ constant, the field mass is given by $\kappa\, e^{-\sigma/2}$, as is seen from (5.16), and hence we identify $\kappa\, e^{-\sigma/2}$ as the field mass in general. It is of interest to note that when the field mass is changing (that is, annihilation or creation is taking place) the last term on the right-hand side of (5.16) provides a compensating term to the ${}^*A_{\alpha\beta}$ field. This is also seen from equation (5.25). If the Lorentz gauge condition is satisfied, then (5.27) reduces to

$$0 = h^{\alpha\nu}\,(\kappa^2\ {}^*A_{\beta\nu} + e^{\sigma}\, E_{\beta\nu})(\nabla_\alpha \sigma).$$

If we are to have $\nabla_\alpha \sigma$ nonzero, then the determinant of the coefficients of $\nabla_\alpha \sigma$ in the above equation must vanish. This can occur only if $\kappa^{-2}\, e^{\sigma}$ is an eigenvalue of the ${}^*A_{\alpha\beta}$ field relative to the $E_{\alpha\beta}$ field, of which the eigenvector is $\nabla_\alpha \sigma$. In this case, we then obtain a well-defined eigenvalue problem for the determination of the function e^{σ}. Viewed the other way, we can have σ nonconstant only at those points in space where $\kappa^{-2}\, e^{\sigma}$ is an eigenvalue of ${}^*A_{\alpha\beta}$ relative to $E_{\alpha\beta}$ with corresponding eigenvector $\nabla_\alpha \sigma$, and hence only at such points will the field mass change by a creation or annihilation process. This conclusion is based on the assumption that the Lorentz gauge condition is satisfied. If this assumption is not made, then the picture becomes more complicated, but can be shown to result in similar conclusions.

We have thus shown that in the case of pure field-field interaction, predictions of creation-annihilation processes are provided by the variant field theory. These predictions are forthcoming only if we go back to the original abstract field space and perform those transformations that leave the affinity of the abstract space invariant, so that creation processes may be viewed as resulting from invariance transformations of the affinity.

Chapter VIII

Fluids and Variant Fields

THE FIBERS of the matter continuum considered in Chapters VI and VII were shown to be governed by the system of equations

$$\mu\, U^{\beta}\, \nabla_{\beta} U^{\lambda} + \rho\, h^{\lambda\gamma}\, A_{\gamma\alpha}\, U^{\alpha} = 0.$$

It is evident from inspection of these equations that there are no direct mechanical interactions, such as a pressure or a stress, and hence the matter continuum thus far considered represents matter that is so rarified that no direct mechanical interaction takes place. In the light of the creation-annihilation representations of field space obtained in Chapter VII, it is of interest to question the state of affairs when direct mechanical interactions are included.

There will be many similarities between the analyses of this chapter and those of the preceding two. To avoid repetition, we assume that all statements and relations in VI-1 have been made here, and proceed from this as a starting point.

1. Elementary Macroscopic Fluids

Our consideration of elementary macroscopic fluids rests on the following two constitutive assumptions.

CONSTITUTIVE ASSUMPTION 1.1. *Let field space be such that Constitutive Assumption* VI-2.1 *is satisfied, so that the matter continuum E_m together with the tensor field $h_{\alpha\beta}$ are well defined and that field space is filled with a system of fibers. In addition, let E_m and p be so chosen that the determinant of $h_{\alpha\beta}$ has the value* -1 *when evaluated in the (u, v, w, p) coördinate system.*

CONSTITUTIVE ASSUMPTION 1.2. *Let there be defined over the 3-dimensional matter continuum E_m two invariable distribution functions $dm \geqq 0$ and de and let there be defined over each fiber a function P; these will be referred to as the affine* *mass and charge distributions and the affine pressure, respectively. If Ω_m is a fiber generated from a point of E_m that is contained in a given domain D of $\mathfrak{F}$, then the action of fluid matter for the domain D is given by*

$$ {}_f\mathcal{I} = \int_{E_m \cap D^*} de \int_{\Omega_m \cap D^*} L^\eta{}_{\eta\alpha}\, dx^\alpha - \int_{D^* \cap E_m} dm \int_{D^* \cap \Omega_m} ds $$

$$ - \int_{D^* \cap E_m} dE_m \int_{D^* \cap \Omega_m} P\, ds. \tag{1.1} $$

The first term on the right-hand side of (1.1) is referred to as the *action of electricity*, and the remaining two terms are referred to as the *action of mass and pressure*, respectively.†

Comparing (1.1) with the action for macroscopic matter given by (VI-2.6), we see that (1.1) is just (VI-2.6) with the added term

$$ - \int_{D^* \cap E_m} dE_m \int_{D^* \cap \Omega_m} P\, ds. $$

Hence we have

$$ {}_f\mathcal{I} = \int_{D^*} \{\rho\, U^\alpha L^\eta{}_{\eta\alpha} - f\,\mu\} - \int_{(E_m \cup \Omega_m) \cap D^*} P\, dE_m\, ds. $$

Now, from $(\text{VI-2.5})_2$ we get

$$ P\, dE_m\, ds = f\, P\, dE_m\, dp, $$

and hence

$$ P\, dE_m\, ds = f\, P\, B\, dV, $$

where B is the Jacobian determinant of (u, v, w, p) with respect to (x^1, x^2, x^3, x^4). Let $h_{\alpha\beta}(x)$ stand for $h_{\alpha\beta}$ when evaluated in the $(x^1, \ldots)$ coördinate system and $h_{\gamma\delta}(u)$ stand for $h_{\alpha\beta}$ when evaluated in the $(u, \ldots)$ coördinate system. From the tensor character of $h_{\alpha\beta}$, we get

$$ h_{\alpha\beta}(x) = h_{\gamma\delta}(u)\, \frac{\partial u^\gamma}{\partial x^\alpha} \frac{\partial u^\delta}{\partial x^\beta}, $$

† An interesting derivation of the variational principle for relativistic perfect fluids from the standpoint of imbedding the Einstein equations in a variational statement is given by A. H. Taub: "General Relativistic Variational Principle for Perfect Fluids," *Phys. Rev.*, **94** (1954), pp. 1468–1470.

and hence on taking determinants we obtain

$$h(x) = h(u)\, B^2.$$

By Constitutive Assumption 1.1, we have $h(u) = -1$, so that

$$B = \sqrt{-h(x)}$$

when evaluated in the $(x^1, \ldots)$ coördinate system. Thus we have

$$P\, dE_m\, ds = f\, P \sqrt{-h}\, dV,$$

from which we obtain

$$_f\mathcal{I} = \int_{D^*} [\rho\, U^\alpha L^\eta{}_{\eta\alpha} - f\,(\mu + \sqrt{-h}\, P)]\, dV.$$

Hence the fluid-matter Lagrangian function is given by

$$_f\mathfrak{L} = \rho\, U^\alpha L^\eta{}_{\eta\alpha} - f\,(\mu + P\sqrt{-h}), \tag{1.2}$$

where

$$f^2 = h_{\alpha\beta}\, U^\alpha\, U^\beta, \qquad U^\alpha = dx^\alpha/dp.$$

In order to obtain the equations of motion of the fluid-matter continuum (equations describing the evolution of the 3-continuum E_m as the fiber parameter p changes) and to extremize the total action with respect to the remaining functions $Q_{(\rho)}$, we must extremize the functional $_f\mathcal{I}$ with respect to the choice of the fiber

$$x^\alpha = x^\alpha(u, v, w, p).$$

As in VI-2, let $\underset{w}{\delta}$ denote the operation of varying the fibers. Since the first two terms in (1.1) are the same as those in (VI-2.6), by (VI-2.11) we have

$$\underset{w}{\delta}\left(\int_{E_m\cap D^*} de \int_{\Omega_m\cap D^*} L^\eta{}_{\eta\alpha}\, dx^\alpha - \int_{E_m\cap D^*} dm \int_{m\cap D^*} ds\right)$$
$$= \int_{D^*}\left\{\frac{\mu}{f}\left[h_{\gamma\beta}\frac{dU^\beta}{dp} + \frac{1}{2}(h_{\gamma\alpha,\beta} + h_{\gamma\beta,\alpha} - h_{\alpha\beta,\gamma})\, U^\alpha\, U^\beta \right.\right.$$
$$\left. - h_{\gamma\beta}\, U^\beta \frac{d}{dp}(\ln f)\right]$$
$$\left. + \rho(L^\eta{}_{\eta\alpha,\gamma} - L^\eta{}_{\eta\gamma,\alpha})\, U^\alpha\right\}\underset{w}{\delta x^\gamma}\, dV.$$

Now, since

$$\underset{w}{\delta}\,(P\,ds) = \left\{(P\,ds)_{,\gamma} - \frac{d}{dp}\,[(P\,ds)_{,U^\gamma}]\right\}\underset{w}{\delta x^\gamma}$$

$$= P\,\underset{w}{\delta}\,ds + \left(P_{,\gamma}\,ds - ds_{,U^\gamma}\,\frac{dP}{dp}\right)\underset{w}{\delta x^\gamma},$$

we have

$$\underset{w}{\delta}\int_{E_m\cap D^*} dE_m \int_{\Omega_m\cap D^*} P\,ds = \int_{E_m\cap D^*} dE_m\,\underset{w}{\delta}\int_{\Omega_m\cap D^*} P\,ds$$

$$= \int_{E_m\cap D^*} dE_m \int_{\Omega_m\cap D^*}\left[P\,\underset{w}{\delta}\,ds + \left(P_{,\gamma} - h_{\gamma\beta}\,U^\beta\,\frac{dP}{dp}\right)\underset{w}{\delta x^\gamma}\,ds\right]$$

$$= \int_{E_m\cap D^*} dE_m \int_{\Omega_m\cap D^*}\left\{\frac{-P}{f}\left[h_{\gamma\beta}\,\frac{dU^\beta}{dp} + \frac{1}{2}\,(h_{\gamma\alpha,\beta} + h_{\gamma\beta,\alpha} - h_{\alpha\beta,\gamma})\,U^\alpha\,U^\beta\right.\right.$$

$$\left. - h_{\gamma\beta}\,U^\beta\,\frac{d}{dp}\ln f\right]$$

$$\left. + f\left[P_{,\gamma} - h_{\gamma\beta}\,U^\beta\,\frac{dP}{dp}\right]\right\}\underset{w}{\delta x^\gamma}\,dp$$

$$= \int_{D^*}\left\{\frac{-P\sqrt{-h}}{f}\left[h_{\gamma\beta}\,\frac{dU^\beta}{dp} + \frac{1}{2}\,(h_{\gamma\alpha,\beta} + h_{\gamma\beta,\alpha} - h_{\alpha\beta,\gamma})\,U^\alpha\,U^\beta\right.\right.$$

$$\left. - h_{\gamma\beta}\,U^\beta\,\frac{d}{dp}\ln f\right]$$

$$\left. + f\sqrt{-h}\left(P_{,\gamma} - h_{\gamma\beta}\,U^\beta\,\frac{dP}{dp}\right)\right\}\underset{w}{\delta x^\gamma}\,dV.$$

Combining these results, we then obtain

$$\underset{w}{\delta}\,{}_f\mathcal{I} = \int_{D^*}\left\{\frac{\mu + \sqrt{-h}\,P}{f}\left[h_{\gamma\beta}\,\frac{dU^\beta}{dp} - \frac{1}{2}\,(h_{\gamma\alpha,\beta} + h_{\gamma\beta,\alpha}\right.\right.$$

$$\left. - h_{\alpha\beta,\gamma})\,U^\alpha\,U^\beta - h_{\gamma\beta}\,\frac{d}{dp}\ln f\right]$$

$$\left. - f\sqrt{-h}\left(P_{,\gamma} - h_{\gamma\beta}\,U^\beta\,\frac{dP}{dp}\right) + \rho\,U^\alpha\,(L^\eta{}_{\eta\alpha,\gamma} - L^\eta{}_{\eta\gamma,\alpha})\right\}\underset{w}{\delta x^\gamma}\,dV.$$

Noting that

$$L^\eta{}_{\eta\alpha,\gamma} - L^\eta{}_{\eta\gamma,\alpha} = A_{\alpha\gamma},$$

and that

$$\Gamma^{\alpha}{}_{\beta\gamma}(h_{\eta\pi}) = \tfrac{1}{2}\, h^{\alpha\eta}\, (h_{\eta\beta,\gamma} + h_{\eta\gamma,\beta} - h_{\beta\gamma,\eta}),$$

on annulling the variation of ${}_f\mathcal{I}$ and multiplying by $h^{\gamma\lambda}$ we have

$$\frac{\mu + \sqrt{-h}\,P}{f}\left[\frac{dU^\lambda}{dp} + \Gamma^{\lambda}{}_{\alpha\beta}(h_{\eta\pi})\, U^\alpha\, U^\beta - U^\lambda \frac{d}{dp}\ln f\right]$$
$$- h^{\gamma\lambda} f \sqrt{-h}\left(P_{,\gamma} - h_{\gamma\beta}\, U^\beta \frac{dP}{dp}\right) + \rho\, A^{\lambda}_{\cdot\alpha}\, U^\alpha = 0. \qquad (1.3)$$

These are the required equations governing the fibers and hence the evolution of the matter continuum E_m with the parameter p.

We shall now show that (1.3) can be reduced to a simpler form. Denote by ∇_β the covariant derivative formed from the affinity $\Gamma^{\alpha}{}_{\beta\gamma}(h_{\eta\pi})$. It is then evident that

$$\frac{dU}{dp} + \Gamma^{\lambda}{}_{\alpha\beta}(h_{\eta\pi})\, U^\alpha\, U^\beta = U^\beta\, \nabla_\beta U^\lambda$$

and

$$\sqrt{-h}\left(P_{,\gamma} - h_{\gamma\beta}\, U^\beta \frac{dP}{dp}\right) = \sqrt{-h}\,(P_{,\gamma} - h_{\gamma\beta}\, U^\beta\, U^\eta\, P_{,\eta})$$
$$= (\delta^\eta_\gamma - h_{\gamma\beta}\, U^\beta\, U^\eta)\sqrt{-h}\,P_{,\eta} = (\delta^\eta_\gamma - h_{\gamma\beta}\, U^\beta\, U^\eta)\,\nabla_\eta(\sqrt{-h}\,P),$$

since ∇_β annuls $h_{\alpha\gamma}$ and $\sqrt{-h}$. Hence (1.3) becomes

$$\frac{\mu + P\sqrt{-h}}{f}\left\{U^\beta\, \nabla_\beta U^\lambda - U^\lambda \frac{d}{dp}\ln f\right\}$$
$$- f\,(h^{\lambda\eta} - U^\lambda\, U^\eta)\,\nabla_\eta(\sqrt{-h}\,P) + \rho\, h^{\gamma\lambda}\, A_{\gamma\alpha}\, U^\alpha = 0,$$

which may be written in the form

$$\frac{\mu + P\sqrt{-h}}{f}\left\{U^\beta\, \nabla_\beta U^\lambda - U^\lambda \frac{d}{dp}\ln f\right\}$$
$$+ (h^{\lambda\eta} - U^\lambda\, U^\eta)\,\{\rho\, A_{\eta\alpha}\, U^\alpha - f\,\nabla_\eta(\sqrt{-h}\,P)\} = 0. \qquad (1.4)$$

This follows from the fact that we have

$$\rho\, h^{\gamma\lambda}\, A_{\gamma\alpha}\, U^\alpha = (h^{\lambda\eta} - U^\lambda\, U^\eta)\,\rho\, A_{\eta\alpha}\, U^\alpha$$

because of the skew-symmetry of $A_{\eta\alpha}$. Multiplying (1.4) by $h_{\gamma\sigma}U^\sigma$ and noting that

$$h_{\lambda\sigma}\, U^\sigma\, U^\beta\, \nabla_\beta U^\lambda = U^\beta\, \nabla_\beta(\tfrac{1}{2}\, h_{\lambda\sigma}\, U^\lambda\, U^\sigma)$$

since ∇_β annuls $h_{\alpha\gamma}$, we have

$$\frac{\mu + P\sqrt{-h}}{f}\left[U^\beta\, \nabla_\beta\,(\tfrac{1}{2}\, h_{\lambda\sigma}\, U^\lambda\, U^\sigma) - h_{\lambda\sigma}\, U^\lambda\, U^\sigma \frac{d}{dp}\ln f\right] + U^\eta\,(1 - h_{\lambda\sigma}\, U^\lambda\, U^\sigma)\,[\rho\, U^\alpha\, A_{\eta\alpha} - f\,\nabla_\eta(\sqrt{-h}\, P)] = 0.$$

Noting that the terms involving $A_{\alpha\beta}$ vanish because of the skew-symmetry of $A_{\alpha\beta}$ and the symmetry of $U^\alpha\, U^\beta$, and that

$$h_{\lambda\sigma}\, U^\lambda\, U^\sigma = f^2,$$

we obtain

$$\frac{\mu + P\sqrt{-h}}{f}\left(U^\beta\, \nabla_\beta \frac{1}{2} f^2 - f^2 \frac{d}{dp}\ln f\right) - U^\eta\,(1 - f^2)\, f\, \nabla_\eta(\sqrt{-h}\, P) = 0,$$

which implies

$$f\,(1 - f^2)\; U^\eta\, \nabla_\eta(\sqrt{-h}\, P) = 0,$$

since

$$U^\beta\, \nabla_\beta \frac{1}{2} f^2 = \frac{d}{dp}\frac{1}{2} f^2 = f \frac{d}{dp} f = f^2 \frac{d}{dp}\ln f.$$

We have thus proved the following result.

THEOREM 1.1. *If $\mathfrak{F}$ is a field space for which Constitutive Assumptions* 1.1 *and* 1.2 *are satisfied, then the equations of motion of the matter fibers are given by*

$$\frac{\mu + P\sqrt{-h}}{f}\left(U^\beta\, \nabla_\beta U^\lambda - U^\lambda \frac{d}{dp}\ln f\right) + (h^{\lambda\eta} - U^\lambda\, U^\eta)\,[\rho\, U^\alpha\, A_{\eta\alpha} - f\,\nabla_\eta(\sqrt{-h}\, P)] = 0, \quad (1.5)$$

which imply that the functions f and P must satisfy the relation

$$f\,(1 - f^2)\; U^\eta\, \nabla_\eta(\sqrt{-h}\, P) = 0. \quad (1.6)$$

There are three ways in which equation (1.6) of Theorem 1.1 can be satisfied:

(a) $f^2 = 1$,

(b) $f = 0$,

(c) $U^\eta \nabla_\eta(\sqrt{-h}\, P) = 0$.

We can obviously set $f \geqq 0$ with no loss of generality since the parameters p and s are at our disposal. With $f = 1$, equation (1.6) is satisfied, and by (VI-2.5) the parameters p and s are equal to within a translation, and hence

$$U^\alpha = \frac{dx^\alpha}{ds}.$$

Equations (1.5) then become

$$(\mu + P\sqrt{-h})\, U^\beta \nabla_\beta U^\lambda + (h^{\lambda\eta} - U^\lambda U^\eta)\,\{\rho\, A_{\eta\alpha}\, U^\alpha - \nabla_\eta(\sqrt{-h}\, P)\} = 0,$$

which have the first integral

$$f^2 = h_{\alpha\beta}\, U^\alpha U^\beta = 1.$$

Case (c) is obviously a special case of (a), since the parameter f is arbitrary in (c) and can be taken to be equal to unity with no loss of generality. If $f = 0$, then equations (1.5) are undefined unless the U^β are such that

$$U^\beta \nabla_\beta U^\lambda = 0$$

holds. We have thus proved the following result.

THEOREM 1.2. *If $\mathfrak{F}$ is a field space for which Constitutive Assumptions* 1.1 *and* 1.2 *are satisfied, then either (a) the equations of motion of the matter fibers are given by*

$$(\mu + P\sqrt{-h})\, U^\beta \nabla_\beta U^\lambda + (h^{\lambda\eta} - U^\lambda U^\eta)\,\{\rho\, A_{\eta\alpha}\, U^\alpha - \nabla_\eta(\sqrt{-h}\, P)\} = 0, \quad (1.7)$$

which have the first integral

$$f^2 = h_{\alpha\beta}\, U^\alpha\, U^\beta = 1, \tag{1.8}$$

where

$$U^\alpha = dx^\alpha/ds;$$

or (*b*) *the field* U^α *is such that*

$$U^\alpha\, U^\beta\, h_{\alpha\beta} = 0, \tag{1.9}$$

in which case we have

$$U^\beta\, \nabla_\beta U^\lambda = 0. \tag{1.10}$$

From this point on, we shall ignore the case

$$U^\alpha\, U^\beta\, h_{\alpha\beta} = f^2 = 0.$$

There are several interesting conclusions that can be drawn from Theorem 1.2. First, the only operations that occur in the equations of the matter fibers are well defined in a metric space with metric tensor $h_{\alpha\beta}$. Hence, although the matter fiber equations are derived in field space, they may be interpreted in a subsidiary physical space of the metric variety.

Second, it is evident from inspection of (1.7) that these equations describe a charged nonviscous fluid with mass density μ, charge density ρ, and pressure P, acted on by an electromagnetic field with field tensor $A_{\alpha\beta}$ and by a gravitational field with field potential tensor $h_{\alpha\beta}$. In the limit as P goes to zero, (1.7) goes to (VI-2.15), and hence we recover the equations of macroscopic matter as the pressure goes to zero, that is, in the limit as the direct mechanical interactions go to zero.

We have been led to equations (1.7), for the motion of the matter fibers and their first integral (1.8), directly from variational considerations over fibers with fiber parameter p rather than s, and have then shown that the resulting equations require $p = s$ in general. If we had not proceeded from the point of view of allowing the fiber parameter p to be different from the value s obtained from the quadratic form

$$ds^2 = h_{\alpha\beta}\, dx^\alpha\, dx^\beta$$

along the fiber, then the first integral (1.8) would not be a unique choice. To see this, we first rewrite (1.7), which results in the case $ds = dp$, in the equivalent form

$$U^{\beta} \nabla_{\beta} U^{\lambda} + (h^{\lambda\eta} - U^{\lambda} U^{\eta}) \left(\frac{\rho}{\mu + P\sqrt{-h}} A_{\eta\alpha} U^{\alpha} - \nabla_{\eta} F \right) = 0, \tag{1.11}$$

where

$$F = \int_{P_0}^{P} \frac{d\xi}{\dfrac{\mu(\xi)}{\sqrt{-h}} + \xi}, \tag{1.12}$$

and where an equation of state $\mu = \mu(P)$ is assumed. Setting

$$f^2 = h_{\alpha\beta} U^{\alpha} U^{\beta}$$

and multiplying (1.11) by $h_{\gamma\lambda} U^{\gamma}$, we have

$$\frac{d}{dp} (\tfrac{1}{2} f^2) - (1 - f^2) \frac{dF}{dp} = 0,$$

which has the two possible solutions

$$f^2 = 1, \quad \frac{dF}{dp} \text{ arbitrary,}$$

and

$$F = \frac{1}{2} \ln \left| \frac{1 - f_0^2}{1 - f^2} \right|,$$

where f_0 is the value of f corresponding to $P = P_0$. The latter solution is eliminated in our development since it does not satisfy (1.6) and hence does not result from the general equation (1.5).

It is specifically to be noted that we can only have $f^2 = 1$ for the macroscopic fluid. It was shown in VI-2 that f^2 is a constant for macroscopic matter and could be set equal to unity with no loss of generality, though other choices were possible. This is not so with the macroscopic fluid, since $f^2 = 1$ is the only constant first integral other than $f^2 = 0$.

2. Explicit Form of the Field Variables

Having obtained a description of macroscopic fluid matter by specifying the fluid-matter Lagrangian, and having obtained the equations of motion for the fibers, we now take up the general fields generated by the axioms of field space.

CONSTITUTIVE ASSUMPTION 2.1. *The field Lagrangian is the scalar density*

$$_{F}\mathfrak{L} = \sqrt{-h}\left(\Lambda + a\, h^{\alpha\beta} R_{\alpha\beta} - \frac{b}{4} A_{\alpha\beta} A_{\gamma\lambda} h^{\alpha\gamma} h^{\beta\lambda}\right), \tag{2.1}$$

where a, b, *and* Λ *are constants and* $h_{\alpha\beta}$ *is the symmetric tensor introduced by Constitutive Assumption* 1.1.

Combining (2.1) and (1.2) yields the total Lagrangian

$$_{T}\mathfrak{L} = {}_{F}\mathfrak{L} + {}_{f}\mathfrak{L}$$

given by

$$_{T}\mathfrak{L} = \sqrt{-h}\left(\Lambda + a\, h^{\alpha\beta} R_{\alpha\beta} - \frac{b}{4} A_{\alpha\beta} A_{\gamma\lambda} h^{\alpha\gamma} h^{\beta\lambda} - f\,P\right)$$
$$- f\,\mu + \rho\, U^{\alpha} L^{\eta}{}_{\eta\alpha}, \tag{2.2}$$

where

$$f^2 = h_{\alpha\beta} U^{\alpha} U^{\beta}.$$

Having the total Lagrangian, we can explicitly evaluate the field variables $H^{\alpha\beta}$, $I^{\alpha\beta}$, S^{α}, $P^{\alpha\beta}{}_{\cdot\cdot\gamma}$, $Z^{(\rho)}$, and $W^{(\rho)\eta}$ in terms of the fundamental variables $A_{\alpha\beta}$ and $R_{\alpha\beta}$ and the subsidiary variables $h_{\alpha\beta}$, U^{α}, ρ, μ, and P. Since we have identified the $h_{\alpha\beta}$ with the first ten $Q_{(\rho)}$ and have obtained the fiber equations corresponding to the remaining $Q_{(\rho)}$, we shall write $Z^{\alpha\beta}$ and $W^{\alpha\beta\eta}$ for the expressions

$$\mathfrak{L}_{,h_{\alpha\beta}} \quad \text{and} \quad \mathfrak{L}_{,h_{\alpha\beta,\gamma}},$$

respectively. It is to be noted that (2.2) is equal to the right-hand side of (VI-4.4) diminished by the term $\sqrt{-h}\,f\,P$. This latter term, however, enters only into the calculation of $Z^{\alpha\beta}$, which is diminished by the term

$$\frac{P\sqrt{-h}}{2} U^{\alpha} U^{\beta} + \frac{\sqrt{-h}}{2} P\, h^{\alpha\beta}.$$

All other equations are the same as in VI-4, and in particular we have

$$I^{\alpha\beta} = a\sqrt{-h}\, h^{\alpha\beta}.$$

Proceeding exactly as in VI-4, we obtain the following result.

THEOREM 2.1. *If $\mathfrak{F}$ is a field space for which Constitutive Assumptions* 1.1, 1.2, *and* 2.1 *are satisfied, then the fields* $A_{\alpha\beta}$, $R_{\alpha\beta}$, $h_{\alpha\beta}$, U^α, ρ, μ, *and* P *are such that they satisfy the system of equations*

$$\left(-\frac{b}{2}\sqrt{-h}\,A^{\alpha\beta}_{\cdot\cdot}\right)_{,\beta} = \frac{3}{10}\,(a\sqrt{-h}\,h^{\alpha\beta})_{;\beta} - \frac{\rho}{2}\,U^\alpha, \tag{2.3}$$

$$(a\sqrt{-h}\,h^{\alpha\beta})_{;\gamma} = \tfrac{2}{5}\,\delta^{(\alpha}_{\gamma}\,(a\sqrt{-h}\,h^{\beta)\eta})_{;\eta}, \tag{2.4}$$

$$\begin{aligned} -a\sqrt{-h}\,(R^{\alpha\beta}_{\cdot\cdot} - \tfrac{1}{2}\,R\,h^{\alpha\beta}) + \frac{\Lambda}{2}\sqrt{-h}\,h^{\alpha\beta} &- \\ - \frac{\mu + P\sqrt{-h}}{2}\,U^\alpha\,U^\beta &- \\ - \frac{P\sqrt{-h}}{2}\,h^{\alpha\beta} + \frac{b}{2}\,(A^{\alpha\eta}_{\cdot\cdot}\,A^{\beta}_{\cdot\eta} - \tfrac{1}{4}\,h^{\alpha\beta}\,A^{\gamma\eta}_{\cdot\cdot}\,A_{\gamma\eta})\sqrt{-h} &= 0, \end{aligned} \tag{2.5}$$

$$\begin{aligned} (\mu + P\sqrt{-h})\,U^\beta\,\nabla_\beta U^\lambda + (h^{\lambda\eta} - U^\lambda\,U^\eta)\,[\rho\,A_{\eta\alpha}\,U^\alpha & \\ - \nabla_\eta(\sqrt{-h}\,P)] &= 0, \end{aligned} \tag{2.6}$$

$$f^2 = h_{\alpha\beta}\,U^\alpha\,U^\beta = 1, \quad U^\alpha = \frac{dx^\alpha}{dp} = \frac{dx^\alpha}{ds}\,f. \tag{2.7}$$

As in VI-4, we have three cases to consider: $I^{\alpha\beta}$ of the nonnull class, $I^{\alpha\beta}$ of the seminull class, and $I^{\alpha\beta}$ of the proper null class. Since the equations of this section are identical to those of Chapter VI with respect to the class determination of the $I^{\alpha\beta}$ field, on combining Theorems VI-4.2 through VI-4.5 we get the following result.

THEOREM 2.2. *Let $\mathfrak{F}$ be a field space for which Constitutive Assumptions* 1.1, 1.2, *and* 2.1 *are satisfied. In the first case,*

$$a \neq 0, \quad I^{\alpha\eta}{}_{:\eta} \neq 0,$$

$\mathfrak{F}$ is a well-based space with base field $h_{\alpha\beta}$; the affine connection is given by

$$L^\alpha{}_{\beta\gamma} = \Gamma^\alpha{}_{\beta\gamma}(h_{\eta\pi}) - \tfrac{1}{4}\,(f_\gamma\,\delta^\alpha_\beta + f_\beta\,\delta^\alpha_\gamma - 3f^\alpha_{\cdot}\,h_{\beta\gamma}) \tag{2.8}$$

in terms of 14 potential-like functions $h_{\alpha\beta}$ and f_γ, which must satisfy the equations (2.3), (2.5), (2.6), *and* (2.7); *and the field equations* (2.4) *are identically and algebraically satisfied by* (2.8). *In addition, we have*

$$h_{\alpha\beta;\gamma} = \tfrac{1}{2}\, h_{\alpha\beta}\, f_\gamma - h_{\gamma(\alpha}\, f_{\beta)}, \tag{2.9}$$

$$R_{\eta\lambda} = J_{\eta\lambda} - \tfrac{3}{4}\, h_{\eta\lambda}\, \nabla_\rho f^\rho_{\cdot} + \tfrac{3}{8}\, f_\eta\, f_\lambda, \tag{2.10}$$

$$A_{\mu\lambda} = f_{[\mu,\lambda]}. \tag{2.11}$$

In the second case,

$$a \neq 0, \quad I^{\alpha\eta}{}_{:\eta} = 0,$$

$\mathfrak{F}$ is a metric space with metric tensor $h_{\alpha\beta}$, so that

$$L^\alpha{}_{\beta\gamma} = \Gamma^\alpha{}_{\beta\gamma}(h_{\eta\pi}).$$

In the third case, $a = 0$, the field equations reduce to $0 = 0$ unless

$$L^\eta{}_{\eta[\beta,\alpha]} \neq 0$$

for some (α, β). If

$$L^\eta{}_{\eta[\beta,\alpha]} \neq 0,$$

the choice of the affinity is not unique and this nonuniqueness requires that the $h_{\alpha\beta}$ field be an assigned field. If f_α is not a gradient vector, then either (2.8) or (V-5.6) may be used as the potential forms for the affinity, and the only surviving field equations are (2.3) and (2.6).

3. Charged, Nongravitating Fluids

Set $a = 0$ and $b = 1$ in (2.2). The $I^{\alpha\beta}$ field then vanishes, so that we have the third case of Theorem 2.2. Hence we obtain the system of equations

$$(\sqrt{-h}\, A^{\alpha\beta}_{\cdot\cdot})_{,\beta} = \rho\, U^\alpha \tag{3.1}$$

and

$$(\mu + P\sqrt{-h})\, U^\beta\, \nabla_\beta U^\lambda + (h^{\lambda\eta} - U^\lambda\, U^\eta)\, \{\rho\, A_{\eta\alpha}\, U^\alpha - \nabla_\eta(\sqrt{-h}\, P)\} = 0, \tag{3.2}$$

together with

$$A_{\alpha\beta} = f_{[\alpha,\beta]}, \tag{3.3}$$

$$S^\alpha = -\tfrac{1}{2}\rho\, U^\alpha, \quad S^\alpha{}_{,\alpha} = 0, \tag{3.4}$$

and

$$h_{\alpha\beta;\gamma} = \tfrac{1}{2} f_\gamma\, h_{\alpha\beta} - h_{\gamma(\alpha} f_{\beta)}, \tag{3.5}$$

for the determination of the functions U^α, f_α, $A_{\alpha\beta}$, ρ, μ, and P. As stated in Theorem 2.2 for this case, the tensor $h_{\alpha\beta}$ is assumed given. Equation (3.5) does not place any restrictions on the tensor $h_{\alpha\beta}$ other than that of differentiability, since (3.5) is identically and algebraically satisfied by (2.8). From this point on, the arguments leading to the proof of Theorem VI-5.1 are directly applicable, and hence we have established the following consequence.

THEOREM 3.1. *If $\mathfrak{F}$ is a field space in which Constitutive Assumptions* 1.1, 1.2, *and* 2.1 *are satisfied, and if $a = 0$, $b = 1$, then the variant field equations yield a system of equations in which only operations occur that are well defined in a subsidiary physical metric space with metric tensor $h_{\alpha\beta}$, and the system of equations is identical with the Maxwell–Lorentz equations for the electromagnetic field in the presence of a charged but not gravitating fluid with mass density μ, charge density ρ, and pressure P.*

It is to be noted that equations (3.4) provide a statement of the conservation of macroscopic charge, but no such statement is forthcoming with respect to the mass or the pressure. This is to be expected, since we have effectively eliminated mass from the problem by requiring the fluid to be nongravitating; that is, the tensor $h_{\alpha\beta}$ is assigned rather than derived.

4. The Uncharged Gravitating Fluid in the Absence of Electromagnetic Fields

Set $b = \rho = 0$ in (2.2) and assume that $a \neq 0$. Under these assumptions, (2.3) implies that

$$\tfrac{3}{10}\,(a\sqrt{-h}\, h^{\alpha\beta})_{;\beta} = \tfrac{3}{10}\, I^{\alpha\beta}{}_{;\beta} = 0.$$

The hypotheses of the second case of Theorem 2.2 are thus satisfied, and hence field space reduces to a metric space with metric tensor $h_{\alpha\beta}$ and affine connection

$$L^\alpha{}_{\beta\gamma} = \Gamma^\alpha{}_{\beta\gamma}.$$

The Maxwellian field tensor $A_{\alpha\beta}$ thus vanishes, so that the conditions

$b = \rho = 0$ and $a \neq 0$ imply the absence of electromagnetic effects. By (2.11), we have $0 = f_{[\alpha,\beta]}$ since $A_{\alpha\beta}$ vanishes, so that we can take $f_\alpha = 0$ with no loss of generality. The only surviving field equations are then the 10 equations

$$J^{\alpha\beta}_{\cdot\cdot} - \frac{1}{2} J\, h^{\alpha\beta} - \frac{\Lambda}{2a}\, h^{\alpha\beta} = -\frac{\mu + P\sqrt{-h}}{2a\sqrt{-h}}\, U^\alpha\, U^\beta - \frac{P}{2a}\, h^{\alpha\beta}, \tag{4.1}$$

and the 4 equations

$$(\mu + P\sqrt{-h})\, U^\beta\, \nabla_\beta U^\lambda - (h^{\lambda\eta} - U^\lambda\, U^\eta)\, \nabla_\eta(\sqrt{-h}\, P) = 0, \tag{4.2}$$

for the determination of the 16 functions of U^α, $h_{\alpha\beta}$, μ, and P. Equations (6.1) and (6.2), however, are just those equations that describe the Einstein gravitation theory in the absence of electromagnetic effects when we identify $-\Lambda/2a$ with the cosmological constant and $1/a$ with the matter-field interaction coefficient. We have thus proved the following theorem.

THEOREM 4.1. *Let $\mathfrak{F}$ be a field space for which Constitutive Assumptions* 1.1, 1.2, *and* 2.1 *are satisfied. If $b = \rho = 0$, then $\mathfrak{F}$ is a metric space with metric tensor $h_{\alpha\beta}$, and the variant field equations reduce to the Einstein gravitation equations for a nonviscous fluid with mass density μ and pressure P when we set $-\Lambda/2a$ equal to the cosmological constant and $1/a$ equal to the matter-field interaction coefficient.*

We have seen that in this case we have 14 equations for the determination of 16 variables. We shall now show that in actuality we have 15 equations. Set

$$G^{\alpha\beta} = J^{\alpha\beta}_{\cdot\cdot} - \tfrac{1}{2} J\, h^{\alpha\beta}; \tag{4.3}$$

then by (II-2.8) we have

$$G^{\alpha\beta}{}_{;\beta} = \nabla_\beta G^{\alpha\beta} = 0, \tag{4.4}$$

since the space under consideration is a metric space by Theorem 4.1. Substituting (4.1) into (4.3) yields

$$aG^{\alpha\beta} = \frac{\Lambda}{2}\, h^{\alpha\beta} - \frac{\mu + P\sqrt{-h}}{2\sqrt{-h}}\, U^\alpha\, U^\beta - \frac{P}{2}\, h^{\alpha\beta},$$

so that by (4.4) we have

$$\nabla_\beta \left(\frac{\mu + P\sqrt{-h}}{\sqrt{-h}}\, U^\alpha\, U^\beta + P\, h^{\alpha\beta} \right) = 0, \tag{4.5}$$

where

$$\nabla_\beta h^{\alpha\beta} = 0$$

since the space has metric tensor $h_{\alpha\beta}$. Expanding (4.5) gives

$$0 = \left(\frac{\mu}{\sqrt{-h}} + P\right) U^\beta\, \nabla_\beta U^\alpha$$

$$+\, U^\alpha\, \nabla_\beta \left[\left(\frac{\mu}{\sqrt{-h}} + P\right) U^\beta\right] + h^{\alpha\beta}\, \nabla_\beta P.$$

Substituting (4.2) into the first term of the above equation yields

$$0 = (2h^{\alpha\beta} - U^\alpha\, U^\beta)\, \nabla_\beta P + U^\alpha\, \nabla_\beta \left[\left(\frac{\mu}{\sqrt{-h}} + P\right) U^\beta\right]. \tag{4.6}$$

If we multiply (4.6) by $h_{\alpha\gamma} U^\gamma$, we then have

$$0 = U^\beta\, (2 - h_{\alpha\gamma}\, U^\alpha\, U^\gamma)\, \nabla_\beta P$$

$$+\, U^\alpha\, U^\gamma\, h_{\alpha\gamma}\, \nabla_\beta \left[\left(\frac{\mu}{\sqrt{-h}} + P\right) U^\beta\right].$$

Since (4.2) admits the first integral

$$h_{\alpha\gamma}\, U^\alpha\, U^\gamma = f^2 = 1, \tag{4.7}$$

this reduces to the desired fifteenth equation

$$0 = U^\beta\, \nabla_\beta P + \nabla_\beta \left[\left(\frac{\mu}{\sqrt{-h}} + P\right) U^\beta\right]. \tag{4.8}$$

It is of interest to note the difference between (4.7) and the customary continuity equation for a noncharged relativistic fluid. Lichnerowicz† gives

† A. Lichnerowicz: *Théories relativistes de la gravitation et de l'électromagnétisme*, Masson, Paris (1955), p. 40, equations 19-4 and 19-5.

$$U^\alpha \nabla_\alpha U^\beta - \frac{(h^{\alpha\beta} - U^\alpha U^\beta)\, \nabla_\alpha P}{\dfrac{\mu}{\sqrt{-h}} + P} = 0 \tag{4.9}$$

as the governing equations of a noncharged relativistic fluid, and

$$\nabla_\alpha \left[\left(\frac{\mu}{\sqrt{-h}} + P \right) U^\alpha \right] - U^\alpha \nabla_\alpha P = 0 \tag{4.10}$$

as the continuity equation (note that Lichnerowicz's ρ is the same as our $\mu/\sqrt{-h}$). Equations (4.9) and (4.2) are easily seen to be identical, and (4.10) and (4.7) are the same except for the sign of the $U^\alpha \nabla_\alpha P$ term. Lichnerowicz's derivation of (4.9) and (4.10) is on a somewhat arbitrary splitting of the divergence of the *assumed* tensor

$$T^{\alpha\beta} = \left(\frac{\mu}{\sqrt{-h}} + P \right) U^\alpha U^\beta - P\, h^{\alpha\beta},$$

whereas in our analysis (4.2) and (4.7), together with the derivation of the momentum-energy tensor

$$aG^{\alpha\beta} = {}'T^{\alpha\beta} \neq T^{\alpha\beta},$$

are all obtained from a single variational principle. Having set the divergence of $T^{\alpha\beta}$ equal to zero, Lichnerowicz then equates $T^{\alpha\beta}$ to $aG^{\alpha\beta}$ and thereby obtains equations corresponding to (4.2). There are thus at least two assumptions involved in the derivation of the Einstein field equations along the lines of Lichnerowicz, in addition to the assumption that the space has a Riemannian metric structure, but our equations are all derived from the same variational statement. We thus feel that our formulation is probably nearer to the truth since it assumes less. Unfortunately, an appeal to the Newtonian approximation does not settle the question, since (4.7) and (4.10) both reduce to the correct continuity equation

$$\nabla_\alpha \left(\frac{\mu}{\sqrt{-h}} U^\alpha \right) = 0$$

in this case.

5. Scaling and Normalization

The field equations (2.3)–(2.7) established in Theorem 2.1 relate a system of field variables defined over the abstract field space under the satisfaction of Constitutive Assumptions 1.1, 1.2, and 2.1. In order to state these equations in a more familiar and recognizable form, it is useful to introduce certain scalings and normalizations of the field variables in the abstract field space $\mathfrak{F}$.

We first rewrite equations (2.3), (2.5), (2.6), (2.7), and (2.11) in the following obvious form:

$$(\sqrt{-h}\, A^{\alpha\beta}_{\cdot\,\cdot})_{,\beta} = -\frac{3a}{2b}\sqrt{-h}\, f^{\alpha}_{\cdot} + \frac{\rho}{b}\, U^{\alpha}, \tag{5.1}$$

$$A_{\alpha\beta} = f_{[\alpha,\beta]}, \tag{5.2}$$

$$(\mu + P\sqrt{-h})\, U_{\beta}\, \nabla_{\beta} U^{\lambda} + (h^{\lambda\eta} - U^{\lambda}\, U^{\eta})\,\{\rho\, A_{\eta\alpha}\, U^{\alpha} - \nabla_{\eta}(\sqrt{-h}\, P)\} = 0, \tag{5.3}$$

$$R_{\alpha\beta} - \frac{1}{2} R\, h_{\alpha\beta} - \frac{\Lambda}{2a} h_{\alpha\beta} = -\frac{\mu + P\sqrt{-h}}{2a\sqrt{-h}}\, U_{\dot{\alpha}}\, U_{\dot{\beta}} - \frac{P\, h_{\alpha\beta}}{2a} + \frac{b}{2a}\left(A_{\alpha\eta}\, A^{\eta}{}_{\beta\cdot} - \tfrac{1}{4}\, h_{\alpha\beta}\, A_{uv}\, A^{uv}_{\cdot\,\cdot}\right), \tag{5.4}$$

$$f^2 = h_{\alpha\beta}\, U^{\alpha}\, U^{\beta} = 1, \quad U^{\alpha} = \frac{dx^{\alpha}}{dp} = \frac{dx^{\alpha}}{ds} f, \tag{5.5}$$

$$R_{\eta\lambda} = J_{\eta\lambda} - \tfrac{3}{4}\, h_{\eta\lambda}\, \nabla_{\rho} f^{\rho}_{\cdot} + \tfrac{3}{8}\, f_{\eta}\, f_{\lambda}, \tag{5.6}$$

in which (2.9) has been used to obtain

$$(\sqrt{-h}\, h^{\alpha\beta})_{;\beta} = \tfrac{5}{2}\sqrt{-h}\, f^{\alpha}_{\cdot},$$

and the (α, β) indices have been lowered in (2.5) by use of the tensor $h_{\eta\pi}$. We have seen in IV-4 that the affine connection $L^{\alpha}{}_{\beta\gamma}$ is invariant under a general class of simultaneous conformal, projective, and gauge transformations; in particular, it is invariant under the conformal transformation (normalization)

$$'h_{\alpha\beta} = \sigma\, h_{\alpha\beta}, \tag{5.7}$$

where σ = constant. (See IV-5, where the conformal transformations on the $I^{\alpha\beta}$ field are shown to be equivalent to (5.7) when the field Lagrangian is a scalar density.) Hence $R_{\alpha\beta}$ and $A_{\alpha\beta}$ are invariant under (5.7). The affine connection $\Gamma^{\alpha}{}_{\beta\gamma}(h_{\eta\pi})$ is also invariant under (5.7) since as a function of $h_{\alpha\beta}$ and $h_{\alpha\beta,\gamma}$ together it is homogeneous of degree zero and σ is a constant. Substituting (5.7) into (5.1)–(5.6) gives, on setting

$$'f^2 = 'h_{\alpha\beta}U^\alpha U^\beta$$

and dropping the primes,

$$(\sqrt{-h}\,A^{\alpha\beta}_{\cdot\cdot})_{,\beta} = -\frac{3a}{2b\,\sigma}\sqrt{-h}\,f^\alpha_\cdot + \frac{\rho}{b}\,U^\alpha, \tag{5.8}$$

$$A_{\alpha\beta} = f_{[\alpha,\beta]}, \tag{5.9}$$

$$\left(\mu + \frac{P\sqrt{-h}}{\sigma^2}\right) U^\beta\,\nabla_\beta U^\lambda + (\sigma\,h^{\lambda\eta} - U^\lambda\,U^\eta)\left[\rho\,A_{\eta\alpha}\,U^\alpha - \nabla_\eta\left(\frac{\sqrt{-h}\,P}{\sigma^2}\right)\right] = 0, \tag{5.10}$$

$$R_{\alpha\beta} - \frac{1}{2}\,R\,h_{\alpha\beta} - \frac{\Lambda}{2a\,\sigma}\,h_{\alpha\beta} = -\frac{\mu + \dfrac{P\sqrt{-h}}{\sigma^2}}{2a\sqrt{-h}}\,U_{\dot\alpha}\,U_{\dot\beta} - \frac{P\,h_{\alpha\beta}}{2a\,\sigma} + \frac{b\,\sigma}{2a}\,(A_{\alpha\eta}\,A_{\beta\cdot}^{\cdot\eta} - \tfrac{1}{4}\,h_{\alpha\beta}\,A_{uv}\,A^{uv}_{\cdot\cdot}), \tag{5.11}$$

$$R_{\alpha\beta} = J_{\alpha\beta} - \tfrac{3}{4}\,h_{\alpha\beta}\,\nabla_\rho f^\rho_\cdot + \tfrac{3}{8}\,f_\alpha\,f_\beta, \tag{5.12}$$

$$f^2 = h_{\alpha\beta}\,U^\alpha\,U^\beta = \sigma, \qquad U^\alpha = \frac{dx^\alpha}{dp} = \frac{dx^\alpha}{ds}\,f. \tag{5.13}$$

Since, by $(5.13)_1$, f^2 is no longer unity, we have

$$\frac{ds}{dp} = f = \sigma^{1/2}.$$

Thus, if we set

$$V^\alpha \overset{\text{def}}{=} \frac{dx^\alpha}{ds}, \tag{5.14}$$

$(5.13)_2$ becomes

$$U^\alpha = \sigma^{1/2}V^\alpha. \tag{5.15}$$

In order to interpret the resulting equations, we find it necessary to take s rather than p as the independent variable, since it is s that has been used in previous theories. By

$$\frac{ds}{dp} = f = \sigma^{1/2}$$

and (5.15), the above equations become

$$(\sqrt{-h}\, A^{\alpha\beta}_{\cdot\cdot})_{,\beta} = -\frac{3a}{2b\,\sigma}\sqrt{-h}\, f^{\alpha}_{\cdot} + \frac{\rho\sqrt{\sigma}}{b} V^{\alpha}, \quad A_{\alpha\beta} = f_{[\alpha,\beta]}, \tag{5.16}$$

$$\left(\mu + \frac{P\sqrt{-h}}{\sigma^2}\right) V^{\beta}\,\nabla_{\beta}V^{\lambda} + (h^{\lambda\eta} - V^{\lambda}\,V^{\eta})\left\{\rho\sqrt{-\sigma}\, V^{\alpha}\, A_{\eta\alpha} - \nabla_{\eta}\left(\frac{\sqrt{-h}\,P}{\sigma^2}\right)\right\} = 0, \tag{5.17}$$

$$R_{\alpha\beta} - \frac{1}{2} h_{\alpha\beta}\, R - \frac{\Lambda}{2a\,\sigma} h_{\alpha\beta} = -\frac{\left(\mu + \frac{P\sqrt{-h}}{\sigma^2}\right)\sigma}{2a\sqrt{-h}} V_{\dot\alpha}\, V_{\dot\beta} - \frac{P\, h_{\alpha\beta}}{2a\,\sigma} + \frac{b\,\sigma}{2a}\left(A_{\alpha\eta}\, A_{\beta\cdot}^{\ \eta} - \tfrac{1}{4} h_{\alpha\beta}\, A_{uv}\, A^{uv}_{\cdot\cdot}\right), \tag{5.18}$$

$$R_{\alpha\beta} = J_{\alpha\beta} - \tfrac{3}{4} h_{\alpha\beta}\, \nabla_{\rho} f^{\rho}_{\cdot} + \tfrac{3}{8} f_{\alpha}\, f_{\beta}, \tag{5.19}$$

$$g^2 = h_{\alpha\beta}\, V^{\alpha}\, V^{\beta} = 1, \quad V^{\alpha} = \frac{dx^{\alpha}}{ds}. \tag{5.20}$$

If we now set

$$\begin{gathered} a = b = 1, \quad f_{\alpha} = \lambda\, \tilde\varphi_{\alpha}, \quad A_{\alpha\beta} = \lambda\, {}^{*}A_{\alpha\beta}, \quad {}^{*}A_{\alpha\beta} = \tilde\varphi_{[\alpha,\beta]}, \\ \tilde\mu = X\,\mu, \quad \tilde\Lambda = -\Lambda/\sigma, \quad \tilde P = X\,P/\sigma^2, \\ \tilde\rho = X\,\rho\,\lambda\sqrt{\sigma}, \quad \lambda^2 = 2\xi\,\kappa^2/3 = \frac{1}{X}, \quad \sigma = \frac{3}{2\kappa^2} = \frac{\xi}{\lambda^2}, \end{gathered} \tag{5.21}$$

after making algebraic simplifications and dropping the tildes, we have

$$(\sqrt{-h}\, {}^{*}A^{\alpha\beta}_{\cdot\cdot})_{,\beta} = -\kappa^2\sqrt{-h}\, \varphi^{\alpha}_{\cdot} + \rho\, V^{\alpha}, \tag{5.22}$$

$$*A_{\alpha\beta} = \varphi_{[\alpha,\beta]}, \tag{5.23}$$

$$(\mu + P\sqrt{-h})\, V^\beta\, \nabla_\beta V^\lambda$$

$$+ (h^{\lambda\eta} - V^\lambda\, V^\eta)\, [\rho\, V^\alpha\, A_{\eta\alpha} - \nabla_\eta(\sqrt{-h}\, P)] = 0, \tag{5.24}$$

$$R_{\alpha\beta} - \frac{1}{2} R\, h_{\alpha\beta} + \frac{\Lambda}{2} h_{\alpha\beta} = -\frac{\xi\,(\mu + P\sqrt{-h})}{2\sqrt{-h}}\, V_{\dot\alpha}\, V_{\dot\beta}$$

$$- \frac{\xi\, P}{2} h_{\alpha\beta} + \frac{\xi}{2} (A_{\alpha\eta}\, A_{\beta\cdot}^{\;\;\eta} - \tfrac{1}{4}\, h_{\alpha\beta}\, A_{\lambda\mu}\, A_{\cdot\cdot}^{\lambda\mu}), \tag{5.25}$$

$$R_{\alpha\beta} = J_{\alpha\beta} + \frac{\xi\,\kappa^2}{4} f_\alpha\, f_\beta - \sqrt{\frac{3\xi}{8}}\, \kappa\, h_{\alpha\beta}\, \nabla_\rho f^\rho_{\cdot}, \tag{5.26}$$

$$g^2 = h_{\alpha\beta}\, V^\alpha\, V^\beta = 1, \quad V^\alpha = \frac{d\kappa^2}{ds}. \tag{5.27}$$

We have thus proved the following theorem.

THEOREM 5.1. *Let*

(a) $\mathfrak{F}$ *be a field space for which Constitutive Assumptions* 1.1, 1.2, *and* 2.1 *are satisfied;*

(b) *the* $h_{\alpha\beta}$ *field be normalized by the conformal transformation*

$$'h_{\alpha\beta} = \sigma\, h_{\alpha\beta}, \tag{5.28}$$

where

$$'f^2 = 'h_{\alpha\beta}\, U^\alpha\, U^\beta$$

and $\sigma =$ *constant, and then delete all primes from the resulting equations;*

(c) p *be replaced as independent fiber parameter by the parameter* s, *which is related to* p *by*

$$\frac{ds}{dp} = \sigma^{1/2}, \tag{5.29}$$

so that

$$U^\alpha = \sigma^{1/2}\, V^\alpha, \tag{5.30}$$

where

$$V^\alpha = dx^\alpha/ds; \tag{5.31}$$

(d) *the following scalings be made:*

$$f_\alpha = \lambda\, \varphi_\alpha, \quad A_{\alpha\beta} = \lambda\, {}^*A_{\alpha\beta}, \quad {}^*A_{\alpha\beta} = \varphi_{[\alpha,\beta]},$$

$$\tilde{\mu} = X\,\mu, \quad \tilde{\Lambda} = \frac{\Lambda}{\sigma}, \quad P = X\,P/\sigma^2, \tag{5.32}$$

$$\tilde{\rho} = X\,\rho\,\lambda\,\sqrt{\sigma},$$

and then delete all tildes from the resulting equations. If we take

$$a = b = 1, \tag{5.33}$$

so that the original field Lagrangian function is independent of arbitrary factors, and take

$$\lambda^2 = \frac{2}{3}\kappa^2\,\xi, \quad \sigma = \frac{3}{2\kappa^2}, \quad X = \frac{1}{\lambda^2}, \tag{5.34}$$

then the field equations (5.1)–(5.6) *assume the form*

$$(\sqrt{-h}\;{}^*A^{\alpha\beta}_{\cdot\cdot})_{,\beta} = -\kappa^2\sqrt{-h}\,\varphi^{\alpha}_{\cdot} + \rho\,V^\alpha, \tag{5.35}$$

$${}^*A_{\alpha\beta} = \varphi_{[\alpha,\beta]}, \quad A_{\alpha\beta} = \kappa\sqrt{\frac{2\xi}{3}}\;{}^*A_{\alpha\beta}, \tag{5.36}$$

$$(\mu + P\sqrt{-h})\,V^\beta\,\nabla_\beta V^\lambda + (h^{\lambda\eta} - V^\lambda\,V^\eta)\,[\rho\,V^\alpha\;{}^*A_{\eta\alpha} - \nabla_\eta(\sqrt{-h}\,P)] = 0, \tag{5.37}$$

$$G_{\alpha\beta} = -\frac{\Lambda}{2} + \frac{\xi\,(\mu + P\sqrt{-h})}{2\sqrt{-h}}\,V_{\dot\alpha}\,V_{\dot\beta} - \frac{\xi\,P}{2}\,h_{\alpha\beta} + \frac{\xi}{2}\,(A_{\alpha\eta}\,A_{\beta\cdot}^{\;\eta} - \tfrac{1}{4}\,h_{\alpha\beta}\,A_{uv}\,A^{uv}_{\cdot\cdot}) - \sqrt{\frac{3\xi}{8}}\,\kappa\,h_{\alpha\beta}\,\nabla_\rho\varphi^\rho_{\cdot} - \frac{\xi\,\kappa^2}{4}\,(\varphi_\alpha\,\varphi_\beta - \tfrac{1}{2}\,\varphi_\rho\,\varphi^\rho_{\cdot}\,h_{\alpha\beta}), \tag{5.38}$$

$$g^2 = h_{\alpha\beta}\,V^\alpha\,V^\beta = 1, \quad V^\alpha = \frac{dx^\alpha}{ds}. \tag{5.39}$$

Here

$$G_{\alpha\beta} = J_{\alpha\beta} - \tfrac{1}{2}\,J\,h_{\alpha\beta} \tag{5.40}$$

is the Einstein tensor, which results from using

$$R_{\alpha\beta} = J_{\alpha\beta} + \frac{\xi\,\kappa^2}{4}\,\varphi_\alpha\,\varphi_\beta - \sqrt{\frac{3\xi}{8}}\,\kappa\,h_{\alpha\beta}\,\nabla_\rho\varphi^{\rho}_{\cdot} \tag{5.41}$$

to evaluate the expression $R_{\alpha\beta} - \frac{1}{2}\,R\,h_{\alpha\beta}$.

By this theorem, we can, with no loss of generality, work with the system of equations (5.33)–(5.40) rather than the system (5.1)–(5.6). Since $a = b = 1$ by (5.33), we see that the constants κ and ξ are completely arbitrary and hence can be chosen at will. Since these constants do not occur in the Lagrangian function, and are not functions of either a or b, they play a role similar to integration constants.

6. Interpretations

We shall now establish interpretations of the various terms in the scaled and normalized field equations (5.35)–(5.40) of Theorem 5.1. In order to establish these interpretations, we rewrite the field equations in such a manner that they involve only processes that are well defined in a subsidiary "physical" metric space.

THEOREM 6.1. *If $\mathfrak{F}$ is a field space for which Constitutive Assumptions* 1.1, 1.2, *and* 2.1 *are satisfied, then there is a distinct subsidiary metric space with metric tensor $h_{\alpha\beta}$ in which the field equations* (5.35)–(5.40) *assume the following form:*

$$\sqrt{-h}\,\nabla_\beta A^{\alpha\beta}_{\cdot\cdot} = -\kappa^2\sqrt{-h}\,\varphi^{\alpha}_{\cdot} + \rho\,V^\alpha, \tag{6.1}$$

$$A_{\alpha\beta} = \nabla_{[\beta}\varphi_{\alpha]}, \tag{6.2}$$

$$(\mu + P\sqrt{-h})\,V^\beta\,\nabla_\beta V^\lambda + (h^{\lambda\eta} - V^\lambda\,V^\eta)\,(\rho\,V^\alpha\,A_{\eta\alpha} - \sqrt{-h}\,\nabla_\eta P) = 0, \tag{6.3}$$

$$\begin{aligned} G_{\alpha\beta} = {} & -\frac{\Lambda}{2}\,h_{\alpha\beta} - \frac{\xi\,(\mu + P\sqrt{-h})}{2\sqrt{-h}}\,V_{\dot\alpha}\,V_{\dot\beta} - \frac{\xi\,P\,h_{\alpha\beta}}{2} \\ & + \frac{\xi}{2}\,(A_{\alpha\eta}\,A_{\beta\cdot}^{\cdot\eta} - \tfrac{1}{4}\,h_{\alpha\beta}\,A_{uv}\,A^{uv}_{\cdot\cdot}) \\ & - \frac{\xi\,\kappa^2}{4}\,(\varphi_\alpha\,\varphi_\beta - \tfrac{1}{2}\,h_{\alpha\beta}\,\varphi_\eta\,\varphi^{\eta}_{\cdot}) - \sqrt{\frac{3\xi}{8}}\,\kappa\,h_{\alpha\beta}\,\nabla_\eta\varphi^{\eta}_{\cdot}, \end{aligned} \tag{6.4}$$

$$g^2 = h_{\alpha\beta} V^\alpha V^\beta = 1, \quad V^\alpha = \frac{dx^\alpha}{ds}, \tag{6.5}$$

where

$$G_{\alpha\beta} = J_{\alpha\beta} - \tfrac{1}{2} J h_{\alpha\beta}. \tag{6.6}$$

Proof. The proof is the same as that of Theorem VII-3.1.

The equations established in Theorem 6.1 are now easily interpretable. We begin, as in VII-3, by establishing the following asymptotic correspondence.

THEOREM 6.2. *If $\mathfrak{F}$ is a field space for which Constitutive Assumptions* 1.1, 1.2, *and* 2.1 *are satisfied, then in the limit as κ approaches zero the solution manifold of the variant field equations over the subsidiary metric space established in Theorem* 6.1 *is isomorphic with the solution manifold of the Einstein–Maxwell theory of interacting gravitational and electromagnetic fields in the presence of a charged fluid with mass density μ, charge density ρ, and pressure P, when we set ξ and Λ equal to the matter-field interaction coefficient and the cosmological constant, respectively.*

Proof. Under the hypotheses, Theorem 6.1 is applicable. Taking the limits of the fluid equations given in Theorem 6.1 as κ goes to zero then gives

$$\sqrt{-h}\, \nabla_\beta A^{\alpha\beta}_{\cdot\cdot} = \rho V^\alpha, \tag{6.6}$$

$$A_{\alpha\beta} = \nabla_{[\beta}\varphi_{\alpha]}, \tag{6.7}$$

$$(\mu + P\sqrt{-h})\, V^\beta \nabla_\beta V^\lambda - (h^{\lambda\eta} - V^\lambda V^\eta)\, \{\rho V^\alpha A_{\eta\alpha} - \sqrt{-h}\, \nabla_\eta P\} = 0, \tag{6.8}$$

$$G_{\alpha\beta} = -\frac{\Lambda}{2} h_{\alpha\beta} - \frac{\xi\,(\mu + P\sqrt{-h})}{2\sqrt{-h}} V_{\dot\alpha} V_{\dot\beta} - \frac{\xi P h_{\alpha\beta}}{2} + 2(A_{\alpha\eta} A_{\beta\cdot}^{\cdot\eta} - \tfrac{1}{4} h_{\alpha\beta} A_{uv} A^{uv}_{\cdot\cdot}), \tag{6.9}$$

$$g^2 = h_{\alpha\beta} V^\alpha V^\beta = 1, \quad V^\alpha = \frac{dx^\alpha}{ds}. \tag{6.10}$$

Since these are just the Einstein–Maxwell equations in the presence of a fluid, the result follows.

Remarks of a nature similar to those following Theorem VII-3.2 are applicable here.

The following theorem, like Theorem VII-3.3, establishes the basis for the interpretation of the general case.

THEOREM 6.3. *If $\mathfrak{F}$ is a field space for which Constitutive Assumptions* 1.1, 1.2, *and* 2.1 *are satisfied, and if $\rho = \mu = P = 0$, then the solution manifold of the variant field equations over the subsidiary metric space established in Theorem* 6.1 *is isomorphic with the solution manifold of the Einstein gravitational field interacting with the (unquantized) vector-meson field with spin* = 1 *and field mass equal to κ, with ξ and Λ set equal to the field-mass interaction coefficient and the cosmological constant, respectively.*

Proof. The proof is the same as that of Theorem VII-3.3.

Combining Theorems 6.2 and 6.3, we make the following interpretation.

If $\mathfrak{F}$ is a field space for which Constitutive Assumptions 1.1, 1.2, *and* 2.1 *are satisfied, then the solution manifold of the variant field equations over the subsidiary metric space established in Theorem* 6.1 *can be interpreted as the solution manifold of the combined Einstein field interacting with a charged macroscopic fluid and the (unquantized) vector-meson field, with field mass κ and matter-field interaction coefficient ξ.*

7. On the Conservation of Mass and Charge

Examining the equations of Theorem 5.1, we see that we have 24 equations, together with the relation implied by

$$S^{\alpha}{}_{,\alpha} = 0, \tag{7.1}$$

and the differential identities of the $J_{\alpha\beta}$ tensor, for the determination of the 27 field variables $A_{\alpha\beta}$, φ_{α}, $h_{\alpha\beta}$, V^{α}, μ, ρ, and P. We now undertake the study of the differential identities of the $J_{\alpha\beta}$ field and the equation (7.1) in order to show that we actually obtain two additional equations, which, together with an equation of state $\mu = \mu(P)$, yield a deterministic system. To facilitate matters, we shall assume throughout this section that the hypotheses of Theorem 5.1 are satisfied and hence equations (5.35)–(5.39) hold.

The first equation to be considered is (7.1). Under the scaling and normalization established in Theorem 5.1, the affine current S^{α} is correspondingly scaled. Since the resulting equations (5.35) are

$$(\sqrt{-h}\ {}^*A^{\alpha\beta})_{,\beta} = -\kappa^2 \sqrt{-h}\ \varphi^{\alpha}_{\cdot} + \rho\ V^{\alpha},$$

which are the same as equations (VII-4.1), the corresponding results of VII-4 hold. We thus have the following analogous result.

THEOREM 7.1. *If the hypotheses of Theorem* 6.1 *hold, then the variant field equations give the following equation for the macroscopic charge density* ρ*:*

$$\nabla_\alpha(\rho\ V^\alpha) = \kappa^2 \sqrt{-h}\ \nabla_\alpha \varphi^{\alpha}_{\cdot}.$$

We now consider the differential identities satisfied by $J_{\alpha\beta}$. By Theorem 6.1, on raising the β index through use of $h^{\beta\gamma}$ and then setting $\gamma = \beta$, we have

$$G^{\beta}{}_{\alpha} = -\frac{\Lambda}{2}\delta^{\beta}_{\alpha} - \frac{\xi}{2\sqrt{-h}}(\mu + P\sqrt{-h})\ V_{\dot\alpha}\ V^{\beta}$$

$$+ \frac{\xi}{2} E^{\beta}{}_{\alpha} - \frac{\xi\,\kappa^2}{4}(\varphi_\alpha\,\varphi^{\beta}_{\cdot} - \tfrac{1}{2}\,\delta^{\beta}_{\alpha}\,\varphi_\rho\,\varphi^{\rho}_{\cdot})$$

$$- \sqrt{\frac{3\xi}{8}}\,\kappa\,\delta^{\beta}_{\alpha}\,\nabla_\rho\varphi^{\rho}_{\cdot} - \frac{\xi}{2}\,\delta^{\beta}_{\alpha}\,P, \tag{7.2}$$

where

$$E^{\beta}{}_{\alpha} = {}^*A_{\alpha\eta}\ {}^*A^{\beta\eta}_{\cdot\cdot} - \tfrac{1}{4}\,\delta^{\beta}_{\alpha}\ {}^*A_{\lambda\eta}\ {}^*A^{\lambda\eta}_{\cdot\cdot}. \tag{7.3}$$

It was shown in II-3 that $J_{\alpha\beta}$ satisfied the differential identity

$$\nabla_\beta G^{\beta}{}_{\alpha} = 0. \tag{7.4}$$

Substituting (7.2) into (7.4) and using the results of VII-4, we get

$$\nabla_\beta G^{\beta}{}_{\alpha} = 0 = -\frac{\xi}{2}\left(\frac{\mu}{\sqrt{-h}} + P\right) V^{\beta}\,\nabla_\beta V_{\dot\alpha}$$

$$- \frac{\xi}{2}\,V_{\dot\alpha}\,\nabla_\beta\left[\left(\frac{\mu}{\sqrt{-h}} + P\right) V^{\beta}\right] - \frac{\xi}{2}\,\nabla_\alpha P$$

$$- \sqrt{\frac{3\xi}{8}}\,\kappa\,\nabla_\alpha\nabla_\rho\varphi^{\rho}_{\cdot} + \frac{\xi}{2}\,\nabla_\beta E^{\beta}{}_{\alpha}$$

$$- \frac{\kappa^2\,\xi}{4}(\varphi_\alpha\,\nabla_\beta\varphi^{\beta}_{\cdot} + \varphi^{\beta}\,\nabla_\beta\varphi_\alpha - \varphi^{\rho}_{\cdot}\,\nabla_\alpha\varphi_\rho). \tag{7.5}$$

Using Lemma VII-4.1 and simplifying, we obtain

$$0 = \left(\frac{\mu}{\sqrt{-h}} + P\right) V^{\beta}\, \nabla_{\beta} V^{\cdot}_{\alpha} + V^{\cdot}_{\alpha}\, \nabla_{\beta}\left[\left(\frac{\mu}{\sqrt{-h}} + P\right) V^{\beta}\right]$$

$$+ \nabla_{\alpha} P + \sqrt{\frac{3}{2\xi}}\, \kappa\, \nabla_{\alpha} \nabla_{\rho} \varphi^{\rho}_{\cdot} + {}^{*}A_{\alpha\eta}\, \nabla_{\beta}\, {}^{*}A^{\eta\beta}_{\cdot\cdot}$$

$$+ \frac{\kappa^2}{2}\,(\varphi_{\alpha}\, \nabla_{\beta} \varphi^{\beta}_{\cdot} + 2\varphi^{\beta}\, \nabla_{[\beta}\varphi_{\alpha]}).$$

By (6.1) and (6.2), we have

$$\nabla_{\beta}\, {}^{*}A^{\eta\beta}_{\cdot\cdot} = -\kappa^2\, \varphi^{\eta}_{\cdot} + \frac{\rho}{\sqrt{-h}}\, V^{\eta}$$

and

$${}^{*}A_{\alpha\beta} = \nabla_{[\beta}\varphi_{\alpha]}.$$

Substituting these into the above equation gives

$$0 = \left(\frac{\mu}{\sqrt{-h}} + P\right) V^{\beta}\, \nabla_{\beta} V^{\cdot}_{\alpha} + V^{\cdot}_{\alpha}\, \nabla_{\beta}\left[\left(\frac{\mu}{\sqrt{-h}} + P\right) V^{\beta}\right]$$

$$+ \nabla_{\alpha} P + \sqrt{\frac{3}{2\xi}}\, \kappa\, \nabla_{\alpha} \nabla_{\rho} \varphi^{\rho}_{\cdot} + {}^{*}A_{\alpha\eta} \left(\frac{\rho}{\sqrt{-h}}\, V^{\eta} - \kappa^2\, \varphi^{\eta}_{\cdot}\right)$$

$$+ \frac{\kappa^2}{2}\,(\varphi_{\alpha}\, \nabla_{\beta} \varphi^{\beta}_{\cdot} + 2\varphi^{\beta}_{\cdot}\, {}^{*}A_{\alpha\beta})$$

$$= \left(\frac{\mu}{\sqrt{-h}} + P\right) V^{\beta}\, \nabla_{\beta} V^{\cdot}_{\alpha} + V^{\cdot}_{\alpha}\, \nabla_{\beta}\left[\left(\frac{\mu}{\sqrt{-h}} + P\right) V^{\beta}\right]$$

$$+ {}^{*}A_{\alpha\eta}\, \frac{\rho}{\sqrt{-h}}\, V^{\eta} + \nabla_{\alpha} P + \kappa\left(\sqrt{\frac{3}{2\xi}}\, \nabla_{\alpha} + \frac{\kappa}{2}\, \varphi_{\alpha}\right) \nabla_{\rho} \varphi^{\rho}_{\cdot}. \qquad (7.6)$$

Now, by (6.3) we have

$$\left(\frac{\mu}{\sqrt{-h}} + P\right) V^{\beta}\, \nabla_{\beta} V^{\cdot}_{\alpha} = -(\delta^{\eta}_{\alpha} - V^{\eta}\, V^{\cdot}_{\alpha}) \left(\frac{\rho}{\sqrt{-h}}\, V^{\nu}\, A_{\eta\nu} - \nabla_{\eta} P\right),$$

and hence (7.6) becomes

$$0 = V^{\cdot}_{\alpha} \nabla_{\beta} \left[\left(\frac{\mu}{\sqrt{-h}} + P \right) V^{\beta} \right] + (2\delta^{\beta}_{\alpha} - V^{\cdot}_{\alpha} V^{\beta}) \nabla_{\beta} P$$

$$+ \kappa \left(\sqrt{\frac{3}{2\xi}} \nabla_{\alpha} + \frac{\kappa}{2} \varphi_{\alpha} \right) \nabla_{\rho} \varphi^{\rho}_{\cdot}.$$

We have thus established the following result.

THEOREM 7.2. *If the hypotheses of Theorem* 6.1 *hold, then the variant field equations give the following system of equations for the macroscopic mass density* μ *and the macroscopic pressure* P:

$$V^{\cdot}_{\alpha} \nabla_{\beta} \left[\left(\frac{\mu}{\sqrt{-h}} + P \right) V^{\beta} \right] + (2\delta^{\beta}_{\alpha} - V^{\cdot}_{\alpha} V^{\beta}) \nabla_{\beta} P$$

$$= -\kappa \left(\sqrt{\frac{3}{2\xi}} \nabla_{\alpha} + \frac{\kappa}{2} \varphi_{\alpha} \right) \nabla_{\rho} \varphi^{\rho}_{\cdot}. \tag{7.7}$$

On elimination of $\nabla_{\rho} \varphi^{\rho}_{\cdot}$, we obtain the following combination of Theorems 7.1 and 7.2.

THEOREM 7.3. *If the hypotheses of Theorem* 6.1 *hold, then the variant field equations imply the conservation of macroscopic charge density* ρ *and macroscopic mass density* μ *if and only if either the Lorentz gauge condition*

$$\nabla_{\sigma} \varphi^{\sigma}_{\cdot} = 0 \tag{7.8}$$

is satisfied or the field mass κ *approaches zero. If neither of these conditions is satisfied then* μ, ρ, *and* P *are related by*

$$V^{\cdot}_{\alpha} \nabla_{\beta} \left[\left(\frac{\mu}{\sqrt{-h}} + P \right) V^{\beta} \right] + (2\delta^{\beta}_{\alpha} - V^{\cdot}_{\alpha} V^{\beta}) \nabla_{\beta} P$$

$$+ \frac{1}{2\kappa} \left(\kappa \varphi_{\alpha} + \sqrt{\frac{6}{\xi}} \nabla_{\alpha} \right) \nabla_{\beta} (\rho V^{\beta}) = 0. \tag{7.9}$$

An alternative form of equation (7.9), which yields a clearer picture of the roles of the various terms, is obtained as follows: Multiply (7.9) by V^{α} to obtain

$$V^\alpha V_{\dot\alpha} \nabla_\beta \left[\left(\frac{\mu}{\sqrt{-h}} + P\right) V^\beta\right] + V^\beta (2 - V^\alpha V_{\dot\alpha}) \nabla_\beta P$$

$$+ \frac{1}{2\kappa}\left(\kappa V^\alpha \varphi_\alpha + \sqrt{\frac{6}{\xi}} V^\alpha \nabla_\alpha\right) \nabla_\beta(\rho V^\beta) = 0.$$

Noting that

$$V^\alpha V_{\dot\alpha} = V^\alpha V^\nu h_{\alpha\nu} = g^2 = 1,$$

we get the required simplified equation,

$$\nabla_\beta \left[\left(\frac{\mu}{\sqrt{-h}} + P\right) V^\beta\right] + V^\beta \nabla_\beta P$$

$$+ \frac{1}{2\kappa}\left(\kappa V^\alpha \varphi_\alpha + \sqrt{\frac{6}{\xi}} V^\alpha \nabla_\alpha\right) \nabla_\beta(\rho V^\beta) = 0. \qquad (7.10)$$

It is now evident from the last three theorems that if we specify an equation of state of the form $\mu = \mu(P)$, then a deterministic system results. In addition, we have shown that results similar to those of Theorems VII-4.1, VII-4.2, and VII-4.3 are also obtained in the case of a macroscopic fluid. Thus the commentary at the end of Chapter VII is applicable here with the trivial additions that must be made to reflect the pressure term P, and hence we obtain annihilation-creation representations in the presence of mechanically interacting matter.

* * *

The results of Part Two are now deemed sufficiently complete to show that the variant field theory, together with certain definitely stated constitutive assumptions, offers an axiomatic foundation for the derivation and analysis of the classical fields. As such, we rest the case of the variant field.

APPENDIXES

Appendix A

Blade Structure, Nonholonomic Coördinates, and Spinor Representations

PARTICULAR INTEREST was shown during the 1940's and 1950's in spinor representations and in the "two-blade" structure of Maxwell–Lorentz electrodynamics. The Dirac electronic equations, as the source of interest in spinor representations, need hardly be discussed. The interest in two-blade structure is principally due to a result that was first discovered by Rainich and later rediscovered by Misner and Wheeler. Consider a 4-dimensional metric space of the Riemannian variety in which Einstein's general-relativity field equations hold. In regions of this space where electromagnetism is the *only* contributor to the stress-energy tensor, and where the electromagnetic field itself is *free* of sources, one can replace the entire content of the combined Einstein–Maxwell theory by a theory that is purely geometrical. It is this result that led Wheeler to what he calls the "already unified" theory,† although the field equations are assumed to be given and then interpreted geometrically rather than being derived from a unified system of purely geometric statements.

The degree of axiomatic constraint for such an already unified theory is quite severe, as is evidenced by the number of conditions that must be satisfied. Another critical feature is that it is not at all evident whether the hypotheses of such a set are independent of each other since they involve both requirements of a geometric nature and requirements springing essentially from physical arguments without any clear separation or distinction between such basically different conditions.

Because of the interest in the Wheeler formalism and for purposes of

† C. W. Misner and J. A. Wheeler: "Classical Physics as Geometry," *Ann. Phys.*, 2 (1957), pp. 525–603. See also J. A. Wheeler: "Neutrinos, Gravitation and Geometry," *Rendiconti della Scuola Internazionale di Fisica "Enrico Fermi,"* Corso XI, Interazioni Deboli (Societá Italiana di Fisica) (1960).

obtaining a set of general results that are useful in studying the physical implications of the variant field theory, we examine the structure of the Maxwellian field tensor $A_{\alpha\beta}$ in this appendix. The examination is based on the eigenvectors and nonholonomic reference frames generated by $A_{\alpha\beta}$, since it is through such considerations that blade structure and spinor analysis are obtained for spaces that are affinely connected but not necessarily metric. The analysis will allow us to demonstrate that in those cases in which the electromagnetic field does not vanish, time-like and space-like directions are always distinguishable in exactly the same way as in the special theory of relativity. One of the basic problems of general relativity is thereby resolved by the variant field theory through inclusion of electromagnetic effects as basic geometric structure. We are thus not faced with the problem of how to interpret results or of how to make space or time measurements in other than weak fields when the variant field theory is used, since the Minkowski structure is evidenced at every point.

A few comments on notation are necessary. With respect to indices, we shall use lower-case Greek letters for tensor indices, lower-case Roman letters at the beginning of the alphabet for matrix or spinor indices, and lower-case Roman letters beginning with j for indices of nonholonomic components of tensors and to denote which eigenvector is under consideration. Capital Roman letters will be used to denote tensors, corresponding lower-case Roman letters to denote tensors referred to nonholonomic frames, and boldface German letters to denote matrices and spinors.

1. Eigenvectors and Eigenvalues of the Maxwellian Tensor $A_{\alpha\beta}$

Let the space considered in this appendix be a field space $\mathfrak{F}$. The customary method of examining the eigenvectors and eigenvalues of a skew-symmetric tensor such as $A_{\alpha\beta}$ requires the use of a symmetric tensor $g_{\alpha\beta}$, which is usually the metric tensor. Since field space in general is not metric, the question of what can be used in place of the metric tensor naturally arises. The tensor $R_{\alpha\beta}$ is well defined and symmetric in field space, and will be used in the ensuing algebraic analysis. When we come to consider differentiation processes, it will be more convenient to replace $R_{\alpha\beta}$ by a symmetric tensor $h_{\alpha\beta}$ that is assumed to be defined over field space and to have a known covariant derivative. The reason for using $R_{\alpha\beta}$, rather than $h_{\alpha\beta}$, for the algebraic analysis is that it allows us to base all of our considerations on curvature expressions.

We first establish certain preliminary results. Let us denote by r the determinant of $R_{\alpha\beta}$ and by a the determinant of $A_{\alpha\beta}$. We explicitly assume that

$$-\infty < r < 0, \qquad a \geqq 0, \tag{1.1}$$

and define the tensor $R^{\alpha\beta}$ by

$$r\, R^{\alpha\beta} = r_{,R_{\alpha\beta}}. \tag{1.2}$$

(The condition $(1.1)_2$ actually is no restriction, since $A_{\alpha\beta}$ is skew-symmetric and real and hence its determinant is a perfect square.) Since $R_{\alpha\beta}$ is a curvature form in $\mathfrak{F}$, the tensor $R^{\alpha\beta}$ can be interpreted geometrically as a "radius of curvature" tensor because of the obvious result that

$$R_{\alpha\beta}\, R^{\beta\gamma} = \delta^{\gamma}_{\alpha} = R^{\gamma\beta}\, R_{\beta\alpha}. \tag{1.3}$$

Let us denote by $e(\lambda)$ the determinant of the tensor

$$E_{\alpha\beta}(\lambda) \stackrel{\text{def}}{=} \lambda\, R_{\alpha\beta} + A_{\alpha\beta}. \tag{1.4}$$

Note. We have

$$E_{\alpha\beta}(-2) = -2R_{\alpha\beta} + A_{\alpha\beta} = -2C_{(\alpha\beta)} - 2C_{[\alpha\beta]} = -2C_{\alpha\beta},$$

since $A_{\alpha\beta} = -2C_{[\alpha\beta]}$ and $R_{\alpha\beta} = C_{(\alpha\beta)}$ by definition (compare II-2). The tensor $E_{\alpha\beta}(\lambda)$ is thus a one-parameter set of curvature expressions that reduces to $-2C_{\alpha\beta}$ for $\lambda = -2$.

It may be shown by direct calculation and use of the skew-symmetry of $A_{\alpha\beta}$ that

$$e(\lambda) = \lambda^4\, r + 2\lambda^2\, A\, r + a, \tag{1.5}$$

where

$$4A \stackrel{\text{def}}{=} A_{\alpha\beta}\, A_{\gamma\sigma}\, R^{\alpha\gamma}\, R^{\beta\sigma}. \tag{1.6}$$

(This result together with a majority of the results of this section are proved by Hlavatý† when we identify $R_{\alpha\beta}$ with Hlavatý's $h_{\alpha\beta}$ and $A_{\alpha\beta}$ with Hlavatý's $k_{\alpha\beta}$.) We refer to a field $A_{\alpha\beta}$ as being of the *first class* if

$$a \neq 0, \qquad A \neq 0, \tag{1.7}$$

† V. Hlavatý: *Geometry of Einstein's Unified Field Theory*, P. Noordhoff, Ltd., Groningen, Holland (1957), Chap. I.

and as being of the *second class* if

$$a = 0, \qquad A \neq 0.\dagger \tag{1.8}$$

A vector U^α will be said to be a *null vector relative* to $R_{\alpha\beta}$ if and only if

$$R_{\alpha\beta}\, U^\alpha\, U^\beta = 0 \tag{1.9}$$

does not imply $U^\alpha = 0$ for all α. Two vectors U^α and V^β will be said to be *perpendicular relative* to $R_{\alpha\beta}$ if and only if the equation

$$R_{\alpha\beta}\, U^\alpha\, V^\beta = 0 \tag{1.10}$$

does not imply either $U^\alpha = 0$ or $V^\alpha = 0$ for all α.

Consider the tensor $A_{\alpha.}^{\ \mu}$ defined by

$$A_{\alpha.}^{\ \mu} = A_{\alpha\beta}\, R^{\beta\mu}. \tag{1.11}$$

A vector U^γ is said to be an *eigenvector* of the tensor $A_{\alpha.}^{\ \eta}$ associated with the *eigenvalue* ρ if and only if U^γ satisfies the equation

$$A_{\alpha.}^{\ \eta}\, U^\alpha = \rho\, U^\eta. \tag{1.12}$$

With these definitions we may establish the following results by direct and elementary algebraic procedures.

(a) If U^α is an eigenvector of $A_{\alpha.}^{\ \eta}$ with nonvanishing eigenvalue, then U^α is a null vector relative to $R_{\alpha\beta}$:

$$R_{\alpha\beta}\, U^\alpha\, U^\beta = 0, \qquad \rho \neq 0. \tag{1.13}$$

(b) If U^α and V^β are two eigenvectors of $A_{\alpha.}^{\ \eta}$ and γ, $\mu \neq \pm\gamma$ are their corresponding eigenvalues, then U^α and V^β are perpendicular relative to $R_{\alpha\beta}$:

$$R_{\alpha\beta}\, U^\alpha\, V^\beta = 0. \tag{1.14}$$

Let $A_{\alpha\beta}$ be of the first class and set

$$D = A^2 - \frac{a}{r}, \tag{1.15}$$

† The case $a = 0$, $A = 0$ is of little interest, and will not be considered. If desired, the analysis can be carried through for this case, but it becomes highly artificial; see Hlavatý: *op. cit.*, pp. 35–40.

which by (1.1) is a nonnegative scalar. The eigenvalues of $A_{\alpha\cdot}^{\ \eta}$ are then given by

$$Y = \underset{1}{\rho} = -\underset{2}{\rho} = i\sqrt{\sqrt{D} + A} \qquad (i = (-1)^{1/2}), \tag{1.16}$$

$$Z = \underset{3}{\rho} = -\underset{4}{\rho} = \sqrt{\sqrt{D} - A}, \tag{1.17}$$

where iY and Z are real. There are only four eigenvectors,

$$\underset{1}{U}^{\alpha}, \qquad \underset{2}{U}^{\alpha}, \qquad \underset{3}{U}^{\alpha}, \qquad \underset{4}{U}^{\alpha},$$

and they have the following properties:

(a) They are defined up to arbitrary factors of proportionality.

(b) They are null vectors relative to $R_{\alpha\beta}$.

(c) The eigenvectors $\underset{1}{U}^{\alpha}$, $\underset{2}{U}^{\alpha}$ are perpendicular, relative to $R_{\alpha\beta}$, to the eigenvectors $\underset{3}{U}^{\alpha}$, $\underset{4}{U}^{\alpha}$.

(d) They satisfy the conditions

$$R_{\alpha\beta}\, \underset{1}{U}^{\alpha}\, \underset{2}{U}^{\beta} \neq 0 \qquad R_{\alpha\beta}\, \underset{3}{U}^{\alpha}\, \underset{4}{U}^{\beta} \neq 0. \tag{1.18}$$

(e) The directions of $\underset{1}{U}^{\alpha}$, $\underset{2}{U}^{\alpha}$ are complex conjugates, and the directions of $\underset{3}{U}^{\alpha}$, $\underset{4}{U}^{\alpha}$ are real.

In addition, the factors of proportionality may be so chosen that the following conditions are satisfied:

(f) The eigenvectors $\underset{1}{U}^{\alpha}$, $\underset{2}{U}^{\alpha}$ are complex conjugates and the eigenvectors $\underset{3}{U}^{\alpha}$, $\underset{4}{U}^{\alpha}$ are real.

(g) We have

$$R_{\alpha\beta}\, \underset{1}{U}^{\alpha}\, \underset{2}{U}^{\beta} = R_{\alpha\beta}\, \underset{3}{U}^{\alpha}\, \underset{4}{U}^{\beta} = 1. \tag{1.19}$$

(h) The eigenvectors satisfying these conditions are uniquely determined to within the transformations

$$'\underset{1}{U}^{\alpha} = e^{i\eta}\, \underset{1}{U}^{\alpha}, \qquad '\underset{2}{U}^{\alpha} = e^{-i\eta}\, \underset{2}{U}^{\alpha}, \tag{1.20}$$

$$'\underset{3}{U}^{\alpha} = \epsilon\, e^{\mu}\, \underset{3}{U}^{\alpha} = '\underset{4}{U}^{A} = \epsilon\, e^{-\mu}\, \underset{4}{U}^{\alpha}, \tag{1.21}$$

where η and μ are arbitrary real scalars and $\epsilon = \pm 1$. Equation (1.20) may be regarded as a rotation in the plane of the real bivector $i\underset{1}{U}^{[\alpha}\underset{2}{U}^{\beta]}$, where η is the angle of rotation and $\underset{1}{U}^{\alpha}$, $\underset{2}{U}^{\alpha}$ are the isotropic vectors of this rotation. (Here $i\underset{1}{U}^{[\alpha}\underset{2}{U}^{\beta]}$ is real since $\underset{1}{U}^{\alpha}$, $\underset{2}{U}^{\alpha}$ are complex conjugates.) On the other hand, (1.21) may be regarded as a Minkowski rotation in the plane of the bivector $\underset{3}{U}^{[\alpha}\underset{4}{U}^{\beta]}$ (followed by a reflection if ϵ has the value -1). The angle of this rotation is μ. It is thus evident that equations (1.20) and (1.21) represent a two-parameter group of motions relative to $R_{\alpha\beta}$, where the group parameters are η and μ. It may also be shown that $i\underset{1}{U}^{[\alpha}\underset{2}{U}^{\beta]}$ and $\underset{3}{U}^{[\alpha}\underset{4}{U}^{\beta]}$ are totally perpendicular planes relative to $R_{\alpha\beta}$. (Every linear combination of $\underset{1}{U}^{\alpha}$ and $\underset{2}{U}^{\alpha}$ is perpendicular, relative to $R_{\alpha\beta}$, to every linear combination of $\underset{3}{U}^{\alpha}$ and $\underset{4}{U}^{\alpha}$.) We thus obtain at every point a system of four vectors exhibiting Lorentz structure, and hence space-like and time-like directions are uniquely determined in field space.

If $A_{\alpha\beta}$ is of the second class, we have the following results:

(a) The eigenvalues of $A_{\alpha\beta}$ are given by

$$\left.\begin{aligned} Y &= \underset{1}{\rho} = -\underset{2}{\rho} = \sqrt{-2A}, \\ Z &= \underset{3}{\rho} = \underset{4}{\rho} = 0, \end{aligned}\right\} \quad A > 0, \tag{1.22}$$

$$\left.\begin{aligned} Y &= \underset{1}{\rho} = \underset{2}{\rho} = 0, \\ Z &= \underset{3}{\rho} = -\underset{4}{\rho} = \sqrt{-2A}, \end{aligned}\right\} \quad A < 0. \tag{1.23}$$

(b) There is a one-parameter family of eigenvectors corresponding to the eigenvalue $\rho = 0$, of which there are only two distinct null vectors relative to $R_{\alpha\beta}$. The directions of these two particular eigenvectors are real [complex conjugate] in the case (1.22) [in the case (1.23)].

(c) The remaining results of the first class apply without change.

2. The Nonholonomic Frame Generated by $A_{\alpha\beta}$

The results presented here hold for both the first-class and the second-class tensors $A_{\alpha\beta}$. In each class, we have a set of four linearly independent eigenvectors† $\underset{j}{U^\alpha}$, which are null vectors relative to $R_{\alpha\beta}$. There is then only one set of linearly independent vectors $\overset{k}{U_\beta}$ satisfying the conditions

$$\underset{j}{U^\alpha}\,\overset{j}{U_\beta} = \delta^\alpha_\beta, \qquad \underset{j}{U^\alpha}\,\overset{k}{U_\alpha} = \delta^k_j. \tag{2.1}$$

We thus obtain both contravariant and covariant eigenvectors of $A_{\alpha\beta}$. Set

$$Z^\alpha{}_j \overset{\text{def}}{=} \underset{j}{U^\alpha}, \qquad Z_\alpha{}^k \overset{\text{def}}{=} \overset{k}{U_\alpha}. \tag{2.2}$$

By (2.1), the functions $Z_\alpha{}^j$ and $Z^\alpha{}_j$ can be looked upon as intermediate components of the unit tensor δ^α_β. With these functions we can define nonholonomic components of any tensorial quantity in $\mathfrak{F}$.

DEFINITION. *The intermediate components $Z_\alpha{}^j$ and $Z^\alpha{}_j$ determine a nonholonomic frame in $\mathfrak{F}$. If $T^{\alpha\cdots}{}_{\beta\ldots}$ are the holonomic components of a tensor in $\mathfrak{F}$, then the corresponding nonholonomic components of $T^{\alpha\cdots}{}_{\beta\ldots}$ are given by*

$$t^{k\cdots}{}_{j\ldots} \overset{\text{def}}{=} Z_\alpha{}^k \cdots Z^\beta{}_j \cdots T^{\alpha\cdots}{}_{\beta\ldots}. \tag{2.3}$$

From this definition and (2.2), it is easily shown that

$$T^{\alpha\cdots}{}_{\beta\ldots} = Z^\alpha{}_j \cdots Z_\beta{}^k \cdots t^{j\cdots}{}_{k\ldots}, \tag{2.4}$$

and that the following results hold:

(a) The nonholonomic components

$$r_{jk} = Z^\alpha{}_j\, Z^\beta{}_k\, R_{\alpha\beta}, \qquad r^{jk} = Z_\alpha{}^j\, Z_\beta{}^k\, R^{\alpha\beta} \tag{2.5}$$

---

† A proof that the $\underset{j}{U^\alpha}$ are linearly independent follows: Set

$$r_{jk} = \underset{j}{U^\alpha}\,\underset{k}{U^\beta}\,R_{\alpha\beta};$$

then we have

$$\det(r_{jk}) = \det(R_{\alpha\beta})\,[\det(\underset{j}{U^\alpha})]^2.$$

One may easily show by (1.13), (1.14), and (1.19) that $r_{jk} = \begin{pmatrix} 0 & 1 & 0 & 0 \\ 1 & 0 & 0 & 0 \\ 0 & 0 & 0 & 1 \\ 0 & 0 & 1 & 0 \end{pmatrix}$, so that $\det(r_{jk}) = 1$. Hence $[\det(\underset{j}{U^\alpha})]^2 = \dfrac{1}{\det(R_{\alpha\beta})} \neq 0$ by (1.1).

are given by

$$((r_{jk})) = ((r^{jk})) = \begin{pmatrix} 0 & 1 & 0 & 0 \\ 1 & 0 & 0 & 0 \\ 0 & 0 & 0 & 1 \\ 0 & 0 & 1 & 0 \end{pmatrix}. \tag{2.6}$$

(b) We have

$$Z^{\alpha}{}_{j} = R^{\alpha\beta}\, r_{kj}\, Z_{\beta}{}^{k}, \qquad Z_{\alpha}{}^{j} = R_{\alpha\beta}\, r^{kj}\, Z^{\beta}{}_{k}, \tag{2.7a}$$

which by (2.6) are equivalent to

$$\begin{aligned} \underset{1}{U^{\alpha}} &= \overset{2}{U^{\alpha}_{\cdot}}, & \underset{2}{U^{\alpha}} &= \overset{1}{U^{\alpha}_{\cdot}}, & \underset{3}{U^{\alpha}} &= \overset{4}{U^{\alpha}_{\cdot}}, & \underset{4}{U^{\alpha}} &= \overset{3}{U^{\alpha}_{\cdot}}, \\ \overset{1}{U_{\alpha}} &= \underset{2}{U_{\dot\alpha}}, & \overset{2}{U_{\alpha}} &= \underset{1}{U_{\dot\alpha}}, & \overset{3}{U_{\alpha}} &= \underset{4}{U_{\dot\alpha}}, & \overset{4}{U_{\alpha}} &= \underset{3}{U_{\dot\alpha}}, \end{aligned} \tag{2.7b}$$

where

$$\overset{j}{U^{\alpha}_{\cdot}} = \overset{j}{U_{\beta}}\, R^{\beta\alpha}, \qquad \underset{j}{U_{\dot\alpha}} = \underset{j}{U^{\beta}}\, R_{\beta\alpha}.$$

(c) The nonholonomic components of $A_{\alpha.}{}^{\eta}$ and $A_{\alpha\beta}$ are given by

$$a_{j.}{}^{k} = \rho_j\, \delta_j^k, \qquad a_{jk} = \rho_j\, r_{jk}, \qquad i \text{ not summed.} \tag{2.8}$$

(d) We have

$$\begin{aligned} R_{\alpha\beta} &= 2\overset{1}{U}_{(\alpha}\, \overset{2}{U}_{\beta)} + 2\overset{3}{U}_{(\alpha}\, \overset{4}{U}_{\beta)}, \\ R^{\alpha\beta} &= 2\underset{1}{U}^{(\alpha}\, \underset{2}{U}^{\beta)} + 2\underset{3}{U}^{(\alpha}\, \underset{4}{U}^{\beta)}, \end{aligned} \tag{2.9}$$

$$A_{\alpha.}{}^{\eta} = \rho_j\, \overset{j}{U_{\alpha}}\, \underset{j}{U^{\beta}}, \tag{2.10}$$

$$A_{\alpha\beta} = 2Y\, \overset{1}{U}_{[\alpha}\, \overset{2}{U}_{\beta]} + 2Z\, \overset{3}{U}_{[\alpha}\, \overset{4}{U}_{\beta]}, \tag{2.11}$$

$$A^{\alpha\beta}_{\cdot\,\cdot} = 2Y\, \underset{2}{U}^{[\alpha}\, \underset{1}{U}^{\beta]} + 2Z\, \underset{4}{U}^{[\alpha}\, \underset{3}{U}^{\beta]}. \tag{2.12}$$

The results established under (a), above, are of central importance in developing spinor representations. In this connection, several observations need to be made. First, if B^α is a real vector, then its first two nonholonomic components b^1 and b^2 are complex conjugates. This follows immediately from the equation

$$b^i = \overset{i}{U}_\alpha \, B^\alpha$$

and the fact that $\overset{1}{U}_\alpha$ and $\overset{2}{U}_\alpha$ are complex conjugates. Let B^α and F_α be two vectors in $\mathfrak{F}$. With any such vectors we may associate the numbers $M(B^\alpha)$ and $M(F_\alpha)$, as defined by the equations

$$M(B^\alpha) \overset{\text{def}}{=} R_{\alpha\beta} \, B^\alpha \, B^\beta, \qquad M(F_\alpha) \overset{\text{def}}{=} R^{\alpha\beta} \, F_\alpha \, F_\beta. \tag{2.13}$$

Since we have taken $R_{\alpha\beta}$ as the fundamental quadratic form in $\mathfrak{F}$, we shall call $M(B^\alpha)$ and $M(F_\alpha)$ the *squares of the moduli of B^α and F_α relative to $R_{\alpha\beta}$ and $R^{\alpha\beta}$, respectively*. Referring these numbers to their nonholonomic evaluation, by (2.6) we have

$$\begin{aligned} M(B^\alpha) &= r_{ij} \, b^i \, b^j = 2(b^1 \, b^2 + b^3 \, b^4), \\ M(F_\alpha) &= 2(f_1 \, f_2 + f_3 \, f_4). \end{aligned} \tag{2.14}$$

Thus B^α $[F_\alpha]$ is a null vector relative to $R_{\alpha\beta}$ (relative to $R^{\alpha\beta}$) if and only if $M(B^\alpha) = 0$ $[M(F_\alpha) = 0]$, or equivalently, if and only if $b^1 \, b^2 = -b^3 \, b^4$ $[f_1 \, f_2 = -f_3 \, f_4]$.

On first examination, the form given by (2.14) may seem a bit strange. In order to cast this in a more recognizable form and explicitly to demonstrate the Lorentz structure of the transformations (1.20) and (1.21), we decompose the nonholonomic components b^j of a real vector B^α by the following transformation:

$$b^j = l^j{}_i \, \sigma^i, \qquad \sigma^i = \overset{-1}{l^i}{}_j \, b^j, \tag{2.15}$$

where

$$\begin{aligned} ((l^j{}_i)) &= \frac{1}{\sqrt{2}} \begin{pmatrix} 1 & i & 0 & 0 \\ 1 & -i & 0 & 0 \\ 0 & 0 & 1 & -1 \\ 0 & 0 & 1 & 1 \end{pmatrix}, \\ ((\overset{-1}{l^j}{}_i)) &= \frac{1}{\sqrt{2}} \begin{pmatrix} 1 & 1 & 0 & 0 \\ -i & i & 0 & 0 \\ 0 & 0 & 1 & 1 \\ 0 & 0 & -1 & 1 \end{pmatrix}. \end{aligned} \tag{2.16}$$

Note. The equations

$$b^1 = \frac{\sigma^1 + i\,\sigma^2}{\sqrt{2}} \quad \text{and} \quad b^2 = \frac{\sigma^1 - i\,\sigma^2}{\sqrt{2}}$$

follow from the complex-conjugate structure of b^1 and b^2.

The factor $1/\sqrt{2}$ is included to simplify the ensuing analysis. Substituting (2.15) and (2.16) into (2.14) gives

$$M(B^\alpha) = (\sigma^1)^2 + (\sigma^2)^2 + (\sigma^3)^2 - (\sigma^4)^2, \tag{2.17}$$

which is immediately recognized as the square of the norm of a vector in a Lorentz space with components σ^k. Under the transformations (1.20) and (1.21) of the eigenvectors and equations (2.15) and (2.16), it can easily be shown that the σ^i transform according to the law

$$\begin{Bmatrix} '\sigma^1 \\ '\sigma^2 \\ '\sigma^3 \\ '\sigma^4 \end{Bmatrix} = \begin{pmatrix} \operatorname{Cos}\eta & \operatorname{Sin}\eta & 0 & 0 \\ -\operatorname{Sin}\eta & \operatorname{Cos}\eta & 0 & 0 \\ 0 & 0 & \epsilon \operatorname{Cosh}\mu & \epsilon \operatorname{Sinh}\mu \\ 0 & 0 & \epsilon \operatorname{Sinh}\mu & \epsilon \operatorname{Cosh}\mu \end{pmatrix} \begin{Bmatrix} \sigma^1 \\ \sigma^2 \\ \sigma^3 \\ \sigma^4 \end{Bmatrix} \tag{2.18}$$

and hence are the components of a vector in a Lorentz space. Equations (2.17) and (2.18) state that the vector space generated by (2.15) and (2.16) from the vector space of the b^i has the same norm and transformation group as a vector space over a Lorentz space. It can also be shown that under (2.15) and (2.16) we have

$$N(B^\alpha, C^\beta) \overset{\text{def}}{=} R_{\alpha\beta}\, B^\alpha\, C^\beta = r_{jk}\, b^j\, c^k = b^1\, c^2 + b^2\, c^1 + b^3\, c^4 + b^4\, c^3$$

$$= \sigma^1\,\lambda^1 + \sigma^2\,\lambda^2 + \sigma^3\,\lambda^3 - \sigma^4\,\lambda^4,$$

where σ^i and λ^i correspond to b^i and c^i under (2.15) and (2.16), respectively. Hence the vector space that results from the vector space of the b^i under (2.15) and (2.16) has the same inner product as a Lorentz vector space. *Thus the vector space generated from the space of b^i by* (2.15) *and* (2.16) *is a Lorentz vector space.*

We now turn to the consideration of the changes in the nonholonomic components of tensors when the eigenvectors undergo the transformations (1.20) and (1.21). Set

$$((c^j{}_k)) = ((c^j{}_k(\eta, \mu))) \stackrel{\text{def}}{=} \begin{pmatrix} e^{-i\eta} & 0 & 0 & 0 \\ 0 & e^{i\eta} & 0 & 0 \\ 0 & 0 & \epsilon e^{-\mu} & 0 \\ 0 & 0 & 0 & \epsilon e^{\mu} \end{pmatrix},$$

$$((\overset{-1}{c^j{}_k})) \stackrel{\text{def}}{=} ((c^j{}_k(-\eta, -\mu))). \tag{2.19}$$

Then we have

$$\operatorname{tr}(c^j{}_k) = \operatorname{tr}(\overset{-1}{c^j{}_k}) = 2\,\mathrm{Cos}\,(\eta) + 2\epsilon\,\mathrm{Cosh}\,(\mu). \tag{2.20}$$

With these quantities, we can rewrite (1.20) and (1.21) as proper transformations:

$$'Z_\alpha{}^j = c^j{}_k\, Z_\alpha{}^k, \qquad 'Z^\alpha{}_j = \overset{-1}{c^k{}_j}\, Z^\alpha{}_k. \tag{2.21}$$

By (2.21) and a direct application of the definition of the nonholonomic components of a tensor, we have the statement:

If the nonholonomic frame is subjected to a transformation of the group defined by (2.19)–(2.21), *the nonholonomic components* $t^{k\cdots}{}_{j\ldots}$ *of a known tensor* $T^{\alpha\cdots}{}_{\beta\ldots}$ *undergo the transformation*

$$'t^{k\cdots}{}_{j\ldots} = \overset{-1}{c^k{}_m} \cdots c^n{}_j \cdots t^{m\cdots}{}_{n\ldots}, \tag{2.22}$$

and the nonholonomic components of $R_{\alpha\beta}$ *and* $A_{\alpha\beta}$ *are invariant.*

The above results allow us to demonstrate the two-blade structure of the affinely connected space $\mathfrak{F}$, as exhibited by the tensor $A_{\alpha\beta}$. From (2.6), $(2.8)_2$, and the fact that $Y = \underset{1}{\rho} = -\underset{2}{\rho}$, $Z = \underset{3}{\rho} = -\underset{4}{\rho}$, we have

$$((a_{jk})) = \begin{pmatrix} 0 & Y & 0 & 0 \\ -Y & 0 & 0 & 0 \\ 0 & 0 & 0 & Z \\ 0 & 0 & -Z & 0 \end{pmatrix}. \tag{2.23}$$

The form of (2.23) shows that a_{jk} has components only in the two biplanes spanned by $i\underset{1}{U}{}^{[\alpha}\underset{2}{U}{}^{\beta]}$ and by $\underset{3}{U}{}^{[\alpha}\underset{4}{U}{}^{\beta]}$. In addition, a_{jk} is invariant under the transformation (2.18). This transformation, however, arose from a rotation in the biplane $i\underset{1}{U}{}^{[\alpha}\underset{2}{U}{}^{\beta]}$ and from a Minkowski rotation in the biplane $\underset{3}{U}^{[\alpha}\underset{4}{U}^{\beta]}$. Hence the tensor a_{jk} exhibits two-blade structure. (It is possible to choose coördinates at every point, say x, y, z, t, such that $A_{\alpha\beta}$ has components in the xy and the zt planes only, of the form exhibited by (2.23), and this result is invariant under a rotation of coördinates in the xy plane and under a Minkowski rotation of coördinates in the zt plane.)

3. Spinor Representations

Let $\mathfrak{F}$ be a field space such that $A_{\alpha\beta}$ is a member of either the first or the second class. (See (1.7) and (1.8) for the definition of these classes.) If $R_{\alpha\beta}$ does not have a negative determinant, replace $R_{\alpha\beta}$ by an appropriate $h_{\alpha\beta}$ that satisfies this condition. Under these assumptions, it was shown in Section 1 that at every point of $\mathfrak{F}$ there exist four linearly independent eigenvectors $\underset{j}{U}^{\alpha}$ of

$$A_{\alpha .}^{\eta} = A_{\alpha\beta}\, R^{\alpha\beta}$$

and a reciprocal set $\overset{j}{U}_{\alpha}$, which are systems of linearly independent null vectors relative to $R_{\alpha\beta}$ and $R^{\alpha\beta}$, respectively. Denote by 0 an arbitrary point of $\mathfrak{F}$ and by $T(0)$ the 4-dimensional tangent space at 0. For the algebraic considerations of this section, we shall require knowledge of quantities in $\mathfrak{F}$ only when evaluated at 0. All quantities in $\mathfrak{F}$ will thus be considered as evaluated at 0 unless explicitly stated to the contrary, and hence may be uniformly defined over $T(0)$.

Since $T(0)$ is a tangent space, we can with no loss of generality consider the eigenvectors $\underset{j}{U}^{\alpha}$ and $\overset{j}{U}_{\alpha}$ as dual systems of base vectors in $T(0)$. With such a system of base vectors over $T(0)$, the transformations (2.19)–(2.21) become proper coördinate transformations in $T(0)$, and the nonholonomic representations of tensors at 0 in $\mathfrak{F}$ become ordinary tensors in $T(0)$ with the transformation rule (2.22).

The basis upon which we shall erect 4-component spinor representations is the following collection of *representation matrices:*

$$(\mathfrak{G}^{a}{}_{b\gamma}) = \mathfrak{G}_{\gamma} \overset{\text{def}}{=} \begin{pmatrix} 0 & 0 & \overset{1}{U}_{\gamma} & \overset{3}{U}_{\gamma} \\ 0 & 0 & \overset{4}{U}_{\gamma} & -\overset{2}{U}_{\gamma} \\ \overset{2}{U}_{\gamma} & \overset{3}{U}_{\gamma} & 0 & 0 \\ \overset{4}{U}_{\gamma} & -\overset{1}{U}_{\gamma} & 0 & 0 \end{pmatrix}, \tag{3.1}$$

$$(\mathfrak{N}^a{}_b{}^\gamma) = \mathfrak{N}^\gamma \stackrel{\text{def}}{=} \begin{pmatrix} 0 & 0 & \underset{2}{U^\gamma} & \underset{4}{U^\gamma} \\ 0 & 0 & \underset{3}{U^\gamma} & -\underset{1}{U^\gamma} \\ \underset{1}{U^\gamma} & \underset{4}{U^\gamma} & 0 & 0 \\ \underset{3}{U^\gamma} & -\underset{2}{U^\gamma} & 0 & 0 \end{pmatrix}, \tag{3.2}$$

where the Roman superscript and subscript denote the row and column, respectively. We first derive certain basic results concerning these matrices. Denote by $\mathfrak{I}$ the 4×4 identity matrix $(\delta^a{}_b)$. A direct calculation based on (2.1), (3.1), and (3.2) easily shows that

$$\mathfrak{G}_\gamma \mathfrak{N}^\gamma = \mathfrak{N}^\gamma \mathfrak{G}_\gamma = 2\mathfrak{I}, \tag{3.3}$$

where the indicated multiplication is matrix multiplication:

$$\mathfrak{N}^\gamma \mathfrak{G}_\gamma = (\mathfrak{N}^\gamma \mathfrak{G}_\gamma)^a{}_c = (\mathfrak{N}^a{}_b{}^\gamma \mathfrak{G}^b{}_{c\gamma}).$$

Similarly, by (2.9), we have

$$2\mathfrak{G}_{(\alpha} \mathfrak{G}_{\beta)} = R_{\alpha\beta}\, \mathfrak{I}, \qquad 2\mathfrak{N}^{(\alpha} \mathfrak{N}^{\beta)} = R^{\alpha\beta}\, \mathfrak{I}. \tag{3.4}$$

Taking the traces of (3.3) and (3.4) gives

$$\begin{aligned} &\operatorname{tr}(\mathfrak{N}^\gamma \mathfrak{G}_\gamma) = 8, \qquad \operatorname{tr}(\mathfrak{G}_{(\alpha} \mathfrak{G}_{\beta)}) = 2R_{\alpha\beta}, \\ &\operatorname{tr}(\mathfrak{N}^{(\alpha} \mathfrak{N}^{\beta)}) = 2R^{\alpha\beta}. \end{aligned} \tag{3.5}$$

From (3.4) and

$$R_{\alpha\beta} R^{\beta\gamma} = \delta^\gamma_\alpha$$

we also have

$$\begin{aligned} \delta^\gamma_\alpha \mathfrak{I} &= 4\mathfrak{G}_{(\alpha} \mathfrak{G}_{\beta)} \mathfrak{N}^{(\beta} \mathfrak{N}^{\gamma)} = 4\mathfrak{N}^{(\gamma} \mathfrak{N}^{\beta)} \mathfrak{G}_{(\beta} \mathfrak{G}_{\alpha)} \\ &= 2R_{\alpha\beta} \mathfrak{N}^{(\beta} \mathfrak{N}^{\gamma)} = 2R^{\gamma\beta} \mathfrak{G}_{(\beta} \mathfrak{G}_{\alpha)}, \end{aligned} \tag{3.6}$$

whence, by (3.3), we obtain

$$\begin{aligned} 2\mathfrak{N}^\gamma \mathfrak{G}_\alpha = \delta^\gamma_\alpha \mathfrak{I} &- \mathfrak{N}^\beta \mathfrak{N}^\gamma \mathfrak{G}_\beta \mathfrak{G}_\alpha - \mathfrak{N}^\gamma \mathfrak{N}^\beta \mathfrak{G}_\alpha \mathfrak{G}_\beta \\ &- \mathfrak{N}^\beta \mathfrak{N}^\gamma \mathfrak{G}_\alpha \mathfrak{G}_\beta. \end{aligned}$$

Since $R_{\alpha\beta}$ $(h_{\alpha\beta})$ has been taken as the fundamental symmetric tensor, (3.4) shows that $\mathfrak{G}_\alpha$ and $\mathfrak{N}^\alpha$ satisfy the same anticommutation relations as do Dirac matrices.† In fact, $\mathfrak{G}_\alpha$ and $\mathfrak{N}^\alpha$ play the same role in the geometry of 4-component spinors as is played by the Pauli matrices in the geometry of 2-component spinors. We may even obtain the Pauli matrices from either $\mathfrak{G}_\alpha$ or $\mathfrak{N}^\alpha$. Define the indices $\dot{1}$ and $\dot{2}$ by $\dot{1} \stackrel{\text{def}}{=} 3,\ \dot{2} \stackrel{\text{def}}{=} 4$, and set

$$\mathfrak{G}_\alpha = \left[\begin{array}{c|c} 0 & ((\mathfrak{G}^a{}_{\dot{b}\alpha})) \\ \hline ((\mathfrak{G}^{\dot{a}}{}_{b\alpha})) & 0 \end{array}\right], \qquad a, b = 1, 2;\ \dot{a}, \dot{b} = \dot{1}, \dot{2}.$$

By (3.1), we have

$$\mathfrak{G}^a{}_{\dot{b}\alpha}\,(\underset{3}{U^\alpha} + \underset{4}{U^\alpha}) = \begin{pmatrix} 0 & 1 \\ 1 & 0 \end{pmatrix} = \boldsymbol{\sigma}_x,$$

$$i\,\mathfrak{G}^a{}_{\dot{b}\alpha}\,(\underset{4}{U^\alpha} - \underset{3}{U^\alpha}) = \begin{pmatrix} 0 & -i \\ i & 0 \end{pmatrix} = \boldsymbol{\sigma}_y,$$

$$\mathfrak{G}^a{}_{\dot{b}\alpha}\,(\underset{2}{U^\alpha} + \underset{1}{U^\alpha}) = \begin{pmatrix} 1 & 0 \\ 0 & -1 \end{pmatrix} = \boldsymbol{\sigma}_z,$$

$$\mathfrak{G}^a{}_{\dot{b}\alpha}\,(\underset{2}{U^\alpha} - \underset{1}{U^\alpha}) = \begin{pmatrix} 1 & 0 \\ 0 & 1 \end{pmatrix} = \boldsymbol{\sigma}_t,$$

with similar results for $\mathfrak{G}^{\dot{a}}{}_{b\alpha}$ and $\mathfrak{N}^\alpha$.

With each vector B^α in $\mathfrak{F}$, we can associate a matrix

$$\mathfrak{B} \stackrel{\text{def}}{=} \mathfrak{G}_\alpha\, B^\alpha \tag{3.7}$$

and with each vector T_α in $\mathfrak{F}$ a matrix

$$\mathfrak{T} \stackrel{\text{def}}{=} \mathfrak{N}^\alpha\, T_\alpha. \tag{3.8}$$

Because of the duality expressed by (3.3)–(3.8) between operations with $(\mathfrak{G}_\alpha, B^\alpha)$ and operations with $(\mathfrak{N}^\alpha, T_\alpha)$, we need only to derive the results implied by (3.7). The analogous results implied by (3.8) are then directly obtained by this duality.

Using (3.1) and the definition of the nonholonomic components of a vector, from (3.7) we obtain

$$(\mathfrak{B}^a{}_b) = \begin{pmatrix} 0 & 0 & b^1 & b^3 \\ 0 & 0 & b^4 & -b^2 \\ b^2 & b^3 & 0 & 0 \\ b^4 & -b^1 & 0 & 0 \end{pmatrix}. \tag{3.9}$$

† S. S. Schweber, H. A. Bethe, and F. de Hoffman: *Mesons and Fields, Volume I, Fields*, Row, Peterson and Co., White Plains, New York (1955), p. 14.

Equation (3.9) shows that $\mathfrak{B}$ is a matrix defined over $T(0)$, which, when given, uniquely determines the vector b^i in $T(0)$ corresponding to B^α in $\mathfrak{F}$ at 0. Equation (3.9) also exhibits the Cartan matrix† structure of $\mathfrak{B}$.

It will now be shown that, given $\mathfrak{B}$, we can recover the vector B^α in a formal manner. From (3.7), we have

$$\mathfrak{B}\,\mathfrak{G}_\beta + \mathfrak{G}_\beta\,\mathfrak{B} = B^\alpha\,(\mathfrak{G}_\alpha\,\mathfrak{G}_\beta + \mathfrak{G}_\beta\,\mathfrak{G}_\alpha)$$

$$= 2\mathfrak{G}_{(\alpha}\,\mathfrak{G}_{\beta)}\,B^\alpha;$$

hence, using $(3.4)_1$, we obtain

$$B^\alpha\,R_{\alpha\beta}\,\mathfrak{J} = (\mathfrak{B}\,\mathfrak{G}_\beta + \mathfrak{G}_\beta\,\mathfrak{B}).$$

Multiplying by $R^{\beta\gamma}$ then gives

$$B^\gamma\,\mathfrak{J} = R^{\beta\gamma}\,(\mathfrak{B}\,\mathfrak{G}_\beta + \mathfrak{G}_\beta\,\mathfrak{B}),$$

so that

$$B^\gamma = \tfrac{1}{4}\,R^{\gamma\beta}\,\mathrm{tr}(\mathfrak{B}\,\mathfrak{G}_\beta + \mathfrak{G}_\beta\,\mathfrak{B}). \tag{3.10}$$

Now, let $\mathfrak{P}$ be any matrix of the form (3.9),

$$\mathfrak{P} = \begin{pmatrix} 0 & 0 & a & c \\ 0 & 0 & d & -b \\ b & c & 0 & 0 \\ d & -a & 0 & 0 \end{pmatrix}, \tag{3.11}$$

where a and b are complex conjugates, and c and d are real, but otherwise are four arbitrary constants. (The reason why a and b are required to be complex conjugates is that the first and second nonholonomic components of a real vector are always complex conjugates since the first and second eigenvectors of $A_\alpha{}^\eta$ are complex conjugates. This requirement ensures that (3.11) is of the same form as (3.9).) Proceeding in the same manner as above, we can define the quantities P^γ by

$$P^\gamma = \tfrac{1}{4}\,R^{\gamma\beta}\,\mathrm{tr}(\mathfrak{P}\,\mathfrak{G}_\beta + \mathfrak{G}_\beta\,\mathfrak{P}), \tag{3.12}$$

from which we can easily show by direct computation that the P^γ are all real and are given by

† E. Cartan: *Leçons sur la théorie des spineurs, I, II,* Actualités Scientifiques et Industrielles, Hermann et Cie., Paris (1938).

$$P^\gamma = a \underset{1}{U^\gamma} + b \underset{2}{U^\gamma} + c \underset{3}{U^\gamma} + d \underset{4}{U^\gamma}. \tag{3.13}$$

The quantities P^γ are thus linear combinations of the four vectors $\underset{j}{U^\gamma}$, so that they constitute the components of a vector in $\mathfrak{F}$ at 0. Hence, *any matrix of the form* (3.11), *where a and b are complex conjugates and c and d are real, defines a vector* P^γ *by* (3.12), *and the quantities* (a, b, c, d) *are the non-holonomic components* (p^1, p^2, p^3, p^4) *of* P^γ.

Let $\mathfrak{B}$ and $\mathfrak{P}$ be matrices associated, by (3.7), with the vectors B^α and P^α, respectively, and set

$$\begin{aligned} N(B^\alpha, P^\alpha) &= R_{\alpha\beta}\, B^\alpha P^\beta = b^1 p^2 + b^2 p^1 + b^3 p^4 + b^4 p^3, \\ M(B^\alpha) &= N(B^\alpha, B^\alpha); \end{aligned} \tag{3.14}$$

then we have

$$N(B^\alpha, P^\alpha)\, \mathfrak{J} = \mathfrak{B}\,\mathfrak{P} + \mathfrak{P}\,\mathfrak{B}, \tag{3.15}$$

so that

$$M(B^\alpha)\, \mathfrak{J} = 2\mathfrak{B}\,\mathfrak{B} \tag{3.16}$$

and

$$\det(\mathfrak{B} - \lambda\, \mathfrak{J}) = (\tfrac{1}{2} M(B^\alpha) - \lambda^2)^2. \tag{3.17}$$

The proof of (3.17) follows by a direct computation based on (3.9) and $(3.14)_2$, and (3.16) is an immediate consequence of $(3.14)_2$ and (3.15). The proof of (3.15) is as follows: From $(3.14)_1$, by (3.4) and (3.7), we have

$$\begin{aligned} N(B^\alpha, P^\alpha)\, \mathfrak{J} &= B^\alpha P^\beta R_{\alpha\beta}\, \mathfrak{J} = 2B^\alpha P^\beta\, \mathfrak{G}_{(\alpha}\, \mathfrak{G}_{\beta)} \\ &= (\mathfrak{G}_\alpha B^\alpha)\, (\mathfrak{G}_\beta P^\beta) + (\mathfrak{G}_\beta P^\beta)\, (\mathfrak{G}_\alpha B^\alpha) \\ &= \mathfrak{B}\,\mathfrak{P} + \mathfrak{P}\,\mathfrak{B}. \end{aligned}$$

Thus far, by use of the eigenvectors $\overset{j}{U}_\alpha$ we have been able to associate with each vector B^α a 4×4 matrix $\mathfrak{B}$, and with every 4×4 matrix of the form (3.9) we have been able to associate a vector B^α. The association of $\mathfrak{B}$ with B^α obviously is not unique, however, since the eigenvectors are determined only to within the two-parameter group of transformations (2.19)–(2.21); that is, to each $'U^j{}_\alpha$ obtained under (2.21) there is a corresponding $'\mathfrak{B}$. The transformations of this two-parameter group, when interpreted in $T(0)$, are proper coördinate transformations corresponding to a rotation in (about) the coördinate biplane (coördinate line)

$$i \underset{1}{U}^{[\alpha} \underset{2}{U}^{\beta]} \qquad (i \overset{1}{U}_{[\alpha} \overset{2}{U}_{\beta]})$$

and to a Minkowski rotation in (about) the coördinate biplane (coördinate line)

$$\underset{3}{U}^{[\alpha} \underset{4}{U}^{\beta]} \qquad (\overset{3}{U}_{[\alpha} \overset{4}{U}_{\beta]})$$

in $T(0)$. We thus wish to determine the relations between the various $\mathfrak{B}$'s associated with a given B^α by the various images of $\overset{j}{U}_\alpha$.

Let $\overset{j}{U}_\alpha$ and $'\overset{j}{U}_\alpha$ be two systems of eigenvectors related by (2.19)–(2.21), and let $\mathfrak{B}$ (etc.) and $'\mathfrak{B}$ (etc.) be the corresponding $\mathfrak{B}$'s (etc.). If $\mathfrak{B}$ and $'\mathfrak{B}$ are both to correspond with B^α, and $\mathfrak{P}$ and $'\mathfrak{P}$ are both to correspond with P^α, we must have

$$N(B^\alpha, P^\alpha)\, \mathfrak{J} = \mathfrak{B}\,\mathfrak{P} + \mathfrak{P}\,\mathfrak{B} = {}'\mathfrak{B}\,{}'\mathfrak{P} + {}'\mathfrak{P}\,{}'\mathfrak{B},$$

and

$$\det(\mathfrak{B} - \lambda\, \mathfrak{J}) = (\tfrac{1}{2} M(B^\alpha) - \lambda^2)^2 = \det({}'\mathfrak{B} - \lambda\, \mathfrak{J})$$

must be an identity in λ, under (3.19)–(3.21). We thus require

$$\mathfrak{B}\,\mathfrak{P} + \mathfrak{P}\,\mathfrak{B} = {}'\mathfrak{B}\,{}'\mathfrak{P} + {}'\mathfrak{P}\,{}'\mathfrak{B} \tag{3.18}$$

and

$$\det(\mathfrak{B} - \lambda\, \mathfrak{J}) = \det({}'\mathfrak{B} - \lambda\, \mathfrak{J}) \qquad \text{for all } \lambda. \tag{3.19}$$

Now, since the $'\overset{j}{U}_\alpha$'s are linear homogeneous functions of the $\overset{j}{U}_\alpha$ under (2.19)–(2.21), and the $\mathfrak{B}$'s are linear homogeneous functions of the $\overset{j}{U}_\alpha$ by (3.1) and (3.7), the $'\mathfrak{B}$'s must be linear homogeneous functions of the $\mathfrak{B}$'s. Thus we have

$$'\mathfrak{B} = \mathfrak{E}\,\mathfrak{B}\,\mathfrak{C}, \qquad '\mathfrak{P} = \mathfrak{H}\,\mathfrak{P}\,\mathfrak{K}, \quad \text{etc.,} \tag{3.20}$$

where $\mathfrak{E}$, $\mathfrak{C}$, $\mathfrak{H}$, $\mathfrak{K}$, etc., are matrix functions of (η, μ) by (2.19). Substituting (3.20) into (3.18) and (3.19) gives

$$\mathfrak{B}\,\mathfrak{P} + \mathfrak{P}\,\mathfrak{B} = \mathfrak{G}\,\mathfrak{B}\,\mathfrak{C}\,\mathfrak{H}\,\mathfrak{P}\,\mathfrak{K} + \mathfrak{H}\,\mathfrak{P}\,\mathfrak{K}\,\mathfrak{G}\,\mathfrak{B}\,\mathfrak{C},$$
$$\det(\mathfrak{B} - \lambda\,\mathfrak{J}) = \det(\mathfrak{G}\,\mathfrak{B}\,\mathfrak{C} - \lambda\,\mathfrak{J}), \tag{3.21}$$
$$\det(\mathfrak{P} - \lambda\,\mathfrak{J}) = \det(\mathfrak{H}\,\mathfrak{P}\,\mathfrak{K} - \lambda\,\mathfrak{J})$$

for all λ. Adding to these conditions the condition that $'\mathfrak{B}$ must also have the form (3.11) in order that it actually represent a real vector, we see that the most general solution is given by

$$\mathfrak{G} = \mathfrak{H} = \mathfrak{C}^{-1} = \mathfrak{K}^{-1}, \tag{3.22}$$

where $\mathfrak{C}$ is a diagonal unimodual matrix. Thus by (3.20) and (3.22) we have

$$'\mathfrak{B} = \mathfrak{C}^{-1}\,\mathfrak{B}\,\mathfrak{C}, \qquad '\mathfrak{P} = \mathfrak{C}^{-1}\,\mathfrak{P}\,\mathfrak{C}. \tag{3.23}$$

Accordingly, the problem reduces to finding $\mathfrak{C}$. By (2.19), (2.22), and (3.9), we have

$$'\mathfrak{B} = \begin{pmatrix} 0 & 0 & b^1\,e^{-i\eta} & \epsilon\,b^3\,e^{-\mu} \\ 0 & 0 & \epsilon\,b^4\,e^{\mu} & -b^2\,e^{i\eta} \\ b^2\,e^{i\eta} & \epsilon\,b^3\,e^{-\mu} & 0 & 0 \\ \epsilon\,b^4\,e^{\mu} & -b^1\,e^{-i\eta} & 0 & 0 \end{pmatrix}. \tag{3.24}$$

Combining (3.23) and (3.24) then leads to

$$\mathfrak{C} = \mathfrak{C}(\Phi) = ((\mathfrak{C}^a{}_b)) = \begin{pmatrix} l_1\,e^{-\Phi} & 0 & 0 & 0 \\ 0 & l_1\,e^{\Phi} & 0 & 0 \\ 0 & 0 & l_2\,e^{-\bar{\Phi}} & 0 \\ 0 & 0 & 0 & l_2\,e^{\bar{\Phi}} \end{pmatrix}, \tag{3.25}$$

$$\mathfrak{C}^{-1} = \mathfrak{C}(-\Phi),$$

$$\operatorname{tr}(\mathfrak{C}) = 2l_1 \operatorname{Cosh} \Phi + 2l_2 \operatorname{Cosh} \bar{\Phi},$$

where

$$2\Phi = \eta + i(\mu + (1 - \epsilon)\,\pi/2),$$
$$2\bar{\Phi} = \eta - i(\mu + (1 - \epsilon)\,\pi/2), \tag{3.26}$$

and where

$$l_1 = \pm 1, \qquad l_2 = \pm 1. \tag{3.27}$$

The results stated in (3.23) and (3.25)–(3.27) establish the desired relation between transformations (2.19)–(2.21) in $T(0)$ and transformations induced in the $\mathfrak{B}$. This relation is not unique, however, but rather is two valued: To every transformation of (2.19)–(2.21) in $T(0)$ there correspond two transformations of $\mathfrak{B}$: one for $l_1l_2 = 1$ and one for $l_1l_2 = -1$. For example, the identity transformation of $T(0)$, given by $\eta = \mu = 1 - \epsilon = 0$, yields $\mathfrak{C} = \mathfrak{I}$ for $l_1l_2 = 1$ and yields

$$\mathfrak{C} = \begin{pmatrix} 1 & 0 & 0 & 0 \\ 0 & 1 & 0 & 0 \\ 0 & 0 & -1 & 0 \\ 0 & 0 & 0 & -1 \end{pmatrix}$$

for $l_1l_2 = -1$. Substituting these back into (3.23) gives $'\mathfrak{B} = \mathfrak{B}$ and $'\mathfrak{B} = -\mathfrak{B}$ as the correspondents of the identity transformation in $T(0)$.

According to the usually accepted definition, an analytic spinor $\mathfrak{X}^{a_1 \cdots a_m}{}_{b_1 \cdots b_n}$ of type (m, n) and order $m + n$ is a collection of quantities that transform according to the law

$$\begin{aligned} &'\mathfrak{X}^{e_1 \cdots e_m}{}_{f_1 \cdots f_n} \\ &= \mathfrak{C}^{-1\,e_1}{}_{a_1} \cdots \mathfrak{C}^{-1\,e_m}{}_{a_m} \mathfrak{C}^{b_1}{}_{f_1} \cdots \mathfrak{C}^{b_n}{}_{f_n} \mathfrak{X}^{a_1 \cdots a_m}{}_{b_1 \cdots b_n} \end{aligned} \tag{3.28}$$

when $T(0)$ undergoes the transformation (2.19)–(2.21). Equation (3.23) shows that $\mathfrak{B}$ is a spinor of type (1, 1) and order 2. It is also evident from

$$'\mathfrak{B} = {'\mathfrak{G}_\alpha} B^\alpha = \mathfrak{C}^{-1}\,\mathfrak{B}\,\mathfrak{C} = \mathfrak{C}^{-1}\,\mathfrak{G}_\alpha\,\mathfrak{C}\,B^\alpha$$

that we have

$$'\mathfrak{G}_\alpha = \mathfrak{C}^{-1}\,\mathfrak{G}_\alpha\,\mathfrak{C},$$

and hence the $\mathfrak{G}_\alpha$ form a vector-ordered collection of spinors of type (1, 1) and order 2. From the method whereby we have derived the spinor transformation properties of the $\mathfrak{B}$'s, there is another interesting interpretation of spinors: Spinors of type (1, 1) and order 2 are equivalence classes of representations of vectors in $\mathfrak{F}$ at 0, which are induced by the collection of all transformations (2.19)–(2.21) in the basis vectors of $T(0)$. This interpretation is evident since the relations between the various representations of B^α by $\mathfrak{B}$, $'\mathfrak{B}$, etc., may be compacted by considering (3.23) to define a relation that obviously is symmetric, reflexive, and transitive and hence an equivalence relation.

As a final check, we verify that (3.10) is invariant under spinor transformations: We have

$$'B^\gamma = \tfrac{1}{4}\, R^{\gamma\beta}\, \mathrm{tr}('\mathfrak{B}\, '\mathfrak{G}_\beta + '\mathfrak{G}_\beta\, '\mathfrak{B}) = \tfrac{1}{4}\, R^{\gamma\beta}\, \mathrm{tr}(\mathfrak{C}^{-1}\, (\mathfrak{B}\, \mathfrak{G}_\beta + \mathfrak{G}_\beta\, \mathfrak{B})\, \mathfrak{C})$$
$$= \tfrac{1}{4}\, R^{\gamma\beta}\, \mathrm{tr}(\mathfrak{B}\, \mathfrak{G}_\beta + \mathfrak{G}_\beta\, \mathfrak{B}) = B^\gamma,$$

since the trace of a matrix is invariant under a similarity transformation.

4. Spinor Connections and Invariant Derivatives

The algebraic considerations of the previous section were based on knowledge of the eigenvectors and other quantities at a single point (0) of field space. In actuality, the contravariant eigenvectors and the covariant eigenvectors form a reciprocal system of vector fields over $\mathfrak{F}$; similarly, spinor quantities form fields over $\mathfrak{F}$ since they are defined at every point of $\mathfrak{F}$ as a result of the freedom in the choice of (0). We shall now examine those properties of spinors that are characterized by neighborhood considerations rather than by point considerations; this subject is referred to as spinor analysis.

Spinor analysis requires a spinor connection for the same reasons that tensor analysis requires an affine connection. A set of 64 quantities

$$\mathfrak{A}_\alpha \stackrel{\text{def}}{=} ((\mathfrak{A}^a{}_{b\alpha})) \tag{4.1}$$

is referred to as a spinor connection provided these quantities transform under (2.19)–(2.21) according to the law

$$'\mathfrak{A}_\alpha = \mathfrak{C}^{-1}\, \mathfrak{A}_\alpha\, \mathfrak{C} + \mathfrak{C}^{-1}{}_{,\alpha}\, \mathfrak{C}, \tag{4.2}$$

where the $\mathfrak{C}$'s are given by (3.23)–(3.26) and where η and μ are now functions of the coördinates in $\mathfrak{F}$.

Our task is to find the spinor connection. To facilitate this task, we explicitly assume that the tensor $R_{\alpha\beta}$ used in the previous sections is replaced by a symmetric tensor $h_{\alpha\beta}$ with a nonvanishing negative determinant. Thus quantities that are orthogonal or null relative to $R_{\alpha\beta}$ become orthogonal or null relative to $h_{\alpha\beta}$.

Let D_α denote the covariant derivative with respect to the following two connections: (a) the connection given by the Christoffel symbols $\Gamma^\alpha{}_{\beta\gamma}(h_{\eta\pi})$, (b) the spinor connection $\mathfrak{A}_\alpha$ we are seeking. Thus, if we have a quantity such as $P^a{}_{b\alpha}{}^\beta$, we put

$$D_\gamma P^a{}_{b\alpha}{}^\beta \stackrel{\text{def}}{=} P^a{}_{b\alpha}{}^\beta{}_{,\gamma} + \Gamma^\beta{}_{\sigma\gamma}\, P^a{}_{b\beta}{}^\sigma - \Gamma^\sigma{}_{\alpha\gamma}\, P^a{}_{b\sigma}{}^\beta$$
$$+ \mathfrak{A}^a{}_{c\gamma}\, P^c{}_{b\alpha}{}^\beta - \mathfrak{A}^c{}_{b\gamma}\, P^a{}_{c\alpha}{}^\beta$$
$$= \nabla_\gamma P^a{}_{b\alpha}{}^\beta + \mathfrak{A}^a{}_{c\gamma}\, P^c{}_{b\alpha}{}^\beta - \mathfrak{A}^c{}_{b\gamma}\, P^a{}_{c\alpha}{}^\beta, \tag{4.3}$$

where ∇_β stands for the covariant derivative formed from the connection $\Gamma^\alpha{}_{\beta\gamma}(h_{\eta\pi})$, with span of operation restricted to Greek indices. In particular, we have

$$D_\gamma h_{\alpha\beta} = \nabla_\gamma h_{\alpha\beta} = 0. \tag{4.4}$$

The spinor connection we are seeking is obtained by requiring that we have

$$D_\beta((\mathfrak{G}^a{}_{b\alpha})) = D_\beta \, \mathfrak{G}_\alpha = 0. \tag{4.5}$$

This is equivalent to requiring that the equations

$$D_\beta \mathfrak{N}^\alpha = 0 \tag{4.6}$$

hold, since

$$\mathfrak{N}^\alpha \, \mathfrak{G}_\alpha = \mathfrak{G}_\alpha \, \mathfrak{N}^\alpha = 2\mathfrak{J}.$$

Note. Since $R_{\alpha\beta}$ has been replaced by $h_{\alpha\beta}$, from $(3.4)_1$ we obtain

$$2\mathfrak{G}_{(\alpha} \, \mathfrak{G}_{\beta)} = h_{\alpha\beta}.$$

Thus $D_\beta \mathfrak{G}_\alpha = 0$ implies and is implied by $D_\lambda h_{\alpha\beta} = 0$. It is for this reason that we have used $\Gamma^\alpha{}_{\beta\gamma}(h_{\eta\pi})$ in the definition of D_γ rather than $L^\alpha{}_{\beta\gamma}$; that is, we have done this to make (4.4) hold.

Expanding (4.5) by use of (4.3), we have

$$0 = D_\beta \mathfrak{G}^a{}_{b\alpha} = \nabla_\beta \mathfrak{G}^a{}_{b\alpha} + \mathfrak{G}^c{}_{b\alpha} \, \mathfrak{A}^a{}_{c\beta} - \mathfrak{G}^a{}_{c\alpha} \, \mathfrak{A}^c{}_{b\beta}$$

$$= \nabla_\beta \mathfrak{G}_\alpha + \mathfrak{A}_\beta \, \mathfrak{G}_\alpha - \mathfrak{G}_\alpha \, \mathfrak{A}_\beta.$$

Thus (4.5) is equivalent to

$$\nabla_\beta \mathfrak{G}_\alpha = \mathfrak{G}_\alpha \, \mathfrak{A}_\beta - \mathfrak{A}_\beta \, \mathfrak{G}_\alpha. \tag{4.7}$$

Multiplying (4.7) by $\mathfrak{N}^\alpha$ on the right and then on the left, and adding and subtracting the results, by (3.3) we get

$$4\mathfrak{A}_\beta = \mathfrak{N}^\alpha \, \nabla_\beta \mathfrak{G}_\alpha - (\nabla_\beta \mathfrak{G}_\alpha) \, \mathfrak{N}^\alpha + \mathfrak{N}^\alpha \, \mathfrak{A}_\beta \, \mathfrak{G}_\alpha + \mathfrak{G}_\alpha \, \mathfrak{A}_\beta \, \mathfrak{N}^\alpha \tag{4.8}$$

and

$$\mathfrak{N}^\alpha \, \nabla_\beta \mathfrak{G}_\alpha + (\nabla_\beta \mathfrak{G}_\alpha) \, \mathfrak{N}^\alpha = \mathfrak{G}_\alpha \, \mathfrak{A}_\beta \, \mathfrak{N}^\alpha - \mathfrak{N}^\alpha \, \mathfrak{A}_\beta \, \mathfrak{G}_\alpha, \tag{4.9}$$

respectively. A direct evaluation based on (3.1) and (3.2) leads to

$$\mathfrak{N}^{\alpha}\,\nabla_{\beta}\mathfrak{G}_{\alpha} + (\nabla_{\beta}\mathfrak{G}_{\alpha})\,\mathfrak{N}^{\alpha} = \underset{j}{U}{}^{\alpha}\,\nabla_{\beta}\overset{j}{U}_{\alpha}\,\mathfrak{I}. \tag{4.10}$$

Now, we have

$$\begin{aligned}\underset{j}{U}{}^{\alpha}\,\nabla_{\beta}\overset{j}{U}_{\alpha} &= \underset{1}{U}{}^{\alpha}\,\nabla_{\beta}\overset{1}{U}_{\alpha} + \underset{2}{U}{}^{\alpha}\,\nabla_{\beta}\overset{2}{U}_{\alpha} + \underset{3}{U}{}^{\alpha}\,\nabla_{\beta}\overset{3}{U}_{\alpha} + \underset{4}{U}{}^{\alpha}\,\nabla_{\beta}\overset{4}{U}_{\alpha}\\ &= \overset{2}{U}{}^{\alpha}_{\cdot}\,\nabla_{\beta}\overset{1}{U}_{\alpha} + \overset{1}{U}{}^{\alpha}\,\nabla_{\beta}\overset{2}{U}_{\alpha} + \overset{4}{U}{}^{\alpha}_{\cdot}\,\nabla_{\beta}\overset{3}{U}_{\alpha} + \overset{3}{U}{}^{\alpha}_{\cdot}\,\nabla_{\beta}\overset{4}{U}_{\alpha}\\ &= \nabla_{\beta}(h^{\gamma\alpha}\,\overset{2}{U}_{\gamma}\,\overset{1}{U}_{\alpha} + h^{\gamma\alpha}\,\overset{4}{U}_{\gamma}\,\overset{3}{U}_{\alpha}),\end{aligned}$$

by (2.7b) and the fact that $\nabla_{\beta}h^{\gamma\alpha} = 0$. By (1.19), the quantity inside the parentheses is equal to the number 2, however, and hence we have

$$\underset{j}{U}{}^{\alpha}\,\nabla_{\beta}\overset{j}{U}_{\alpha} = 0.$$

Thus, by (4.9) and (4.10), we get

$$\mathfrak{G}_{\alpha}\,\mathfrak{A}_{\beta}\,\mathfrak{N}^{\alpha} = \mathfrak{N}^{\alpha}\,\mathfrak{A}_{\beta}\,\mathfrak{G}_{\alpha}. \tag{4.11}$$

Equation (4.11) thus reduces (4.8) to the following form:

$$4\mathfrak{A}_{\beta} = \mathfrak{N}^{\alpha}\,\nabla_{\beta}\mathfrak{G}_{\alpha} - (\nabla_{\beta}\mathfrak{G}_{\alpha})\,\mathfrak{N}^{\alpha} + 2\mathfrak{N}^{\alpha}\,\mathfrak{A}_{\beta}\,\mathfrak{G}_{\alpha}. \tag{4.12}$$

Since (4.12) is a linear inhomogeneous equation for the determination of $\mathfrak{A}_{\beta}$, the general solution is given by the sum of a particular solution of the inhomogeneous equation and the general solution of the corresponding homogeneous equation. The homogeneous equation corresponding to (4.12) is

$$2\mathfrak{S}_{\beta} = \mathfrak{N}^{\alpha}\,\mathfrak{S}_{\beta}\,\mathfrak{G}_{\alpha}, \tag{4.13}$$

which implies and is implied by

$$\mathfrak{G}_{\alpha}\,\mathfrak{S}_{\beta} = \mathfrak{S}_{\beta}\,\mathfrak{G}_{\alpha}. \tag{4.14}$$

(To see that (4.13) is implied by (4.14), we need only to multiply (4.14) on the left by $\mathfrak{N}^{\alpha}$ and remember that $\mathfrak{N}^{\alpha}\,\mathfrak{G}_{\alpha} = 2\mathfrak{I}$ to recover (4.13). That (4.14) is implied by the above considerations can be seen directly from (4.7) since (4.14) is just the homogeneous part of (4.7).) The general solution to the homogeneous equation corresponding to (4.12) is therefore given by a collection of matrices $\mathfrak{S}_{\beta}$ that commute with the matrices $\mathfrak{G}_{\alpha}$

for all (α, β), and hence satisfy (4.11). An obvious possible solution to (4.14) is

$$\mathfrak{S}_\beta = W_\beta \mathfrak{J},$$

where W_β is an arbitrary vector. Such an $\mathfrak{S}_\beta$ has the same transformation properties as $\mathfrak{G}_\alpha$.

We are now left with the problem of finding a particular solution to (4.12). Set

$$\mathfrak{M}_\beta = \tfrac{1}{4}\,[\mathfrak{N}^\alpha \nabla_\beta \mathfrak{G}_\alpha - (\nabla_\beta \mathfrak{G}_\alpha)\,\mathfrak{N}^\alpha]. \tag{4.15}$$

A direct computation based on (3.1) and (3.2) gives

$$\mathfrak{M}_\gamma = \left(\begin{pmatrix} X_\gamma & Y_\gamma & 0 & 0 \\ Z_\gamma & -X_\gamma & 0 & 0 \\ 0 & 0 & \bar{X}_\gamma & \bar{Y}_\gamma \\ 0 & 0 & \bar{Z}_\gamma & -\bar{X}_\gamma \end{pmatrix}\right), \tag{4.16}$$

where

$$4X_\gamma = \underset{2}{U^\alpha} \nabla_\gamma \overset{2}{U}_\alpha + \underset{4}{U^\alpha} \nabla_\gamma \overset{4}{U}_\alpha - \underset{1}{U^\alpha} \nabla_\gamma \overset{1}{U}_\alpha - \underset{3}{U^\alpha} \nabla_\gamma \overset{3}{U}_\alpha, \tag{4.17}$$

$$2Y_\gamma = \underset{2}{U^\alpha} \nabla_\gamma \overset{3}{U}_\alpha - \underset{4}{U^\alpha} \nabla_\gamma \overset{1}{U}_\alpha, \tag{4.18}$$

$$2Z_\gamma = \underset{3}{U^\alpha} \nabla_\gamma \overset{2}{U}_\alpha - \underset{1}{U^\alpha} \nabla_\gamma \overset{4}{U}_\alpha, \tag{4.19}$$

and where the bar denotes complex conjugate (remember that the first and second eigenvectors are complex conjugates and the third and fourth are real). After a lengthy but straightforward calculation based on the above equations and (2.1) and (2.2), it can be shown that $\mathfrak{M}$ identically satisfies (4.12), so that we have $\mathfrak{N}^\alpha\,\mathfrak{M}_\beta\,\mathfrak{G}_\alpha = 0$, which implies the satisfaction of (4.11). Hence $\mathfrak{M}$ is the desired particular solution. Thus the general connection determined by (4.5) is given by

$$\mathfrak{A}_\beta = \mathfrak{M}_\beta + \mathfrak{S}_\beta \tag{4.20}$$

for any collection of matrices $\mathfrak{S}_\beta$ that commute with $\mathfrak{G}_\alpha$ for all (α, β). Taking $\mathfrak{S}_\beta = W_\beta\,\mathfrak{J}$, we can then use W_β to adjust the trace of $\mathfrak{A}_\beta$ to any preassigned vector:

$$\mathrm{tr}(\mathfrak{A}_\beta) = \mathrm{tr}(\mathfrak{M}_\beta) + 4W_\beta.$$

It now remains to show that (4.20) transforms according to the law (4.2). Since, by (4.16)–(4.19), $\mathfrak{M}_\gamma$ is a function of the eigenvectors, under the transformations (2.19)–(2.21) of these eigenvectors we have, at (0),

$$'X_\gamma = X_\gamma + \tfrac{1}{2}(\eta + i\,\mu)_{,\gamma}, \qquad 'Y_\gamma = \epsilon\, e^{-(\eta+i\,\mu)}\, Y_\gamma, \qquad 'Z_\gamma = \epsilon\, e^{(\eta+i\,\mu)}\, Z_\gamma. \tag{4.21}$$

By (3.25)–(3.27) and (4.16)–(4.21), it is easily shown that $\mathfrak{M}_\gamma$ transforms according to the law (4.2). *Thus if* $\mathfrak{S}_\gamma$ *transforms according to the spinor law of transformation* (3.23), *the quantities given by* (4.20) *have the required transformation properties of a spinor connection. Hence* (4.20) *is the required spinor connection determined by* (4.5).

We shall now show the convenience of taking (4.5) as the generating equation for the spinor connection. Let $\mathfrak{B}$ be the spinor associated with a vector field B^α in $\mathfrak{F}$ by (3.7), so that we have

$$\mathfrak{B} = \mathfrak{G}_\alpha B^\alpha. \tag{4.22}$$

Computing the D_γ derivative of both sides gives

$$D_\gamma\mathfrak{B} = D_\gamma(\mathfrak{G}_\alpha\, B^\alpha) = (D_\gamma\mathfrak{G}_\alpha)\, B^\alpha + \mathfrak{G}_\alpha\, D_\gamma B^\alpha,$$

so that by (4.5) we obtain

$$D_\gamma\mathfrak{B} = \mathfrak{G}_\alpha\, D_\gamma B^\alpha = \mathfrak{G}_\alpha\, \nabla_\gamma B^\alpha. \tag{4.23}$$

Thus the D_γ derivative of an associated spinor is the spinor associated with the covariant derivative of the generating vector with respect to the connection $\Gamma^\alpha{}_{\beta\gamma}(h_{\eta\pi})$. It is also evident from (4.23) that $P^\gamma\, D_\gamma\mathfrak{B}$ is again a spinor of the same type as $\mathfrak{B}$, where P^γ is an arbitrary nonzero vector, and hence D_γ *differentiation is an invariant process.*

The spaces that have arisen most often in our study of the variant field equations are well-based spaces. In such spaces there is defined a symmetric tensor $h_{\alpha\beta}$ with negative, nonzero determinant, and there is defined a collection of 40 functions $Q_{\gamma(\alpha\beta)} = Q_{\gamma\alpha\beta}$ such that the equation

$$h_{\alpha\beta;\gamma} = -Q_{\gamma\alpha\beta} \tag{4.24}$$

holds at all points. The development leading to the spinor connection (4.20), however, arose from the use of the D_γ derivative and a symmetric tensor $h_{\alpha\beta}$ with negative, nonzero determinant. From the definition of the D_γ derivative, we have

$$D_\gamma h_{\alpha\beta} = \nabla_\gamma h_{\alpha\beta} = 0 \tag{4.25}$$

(see equations (4.3) and (4.4)), and hence the spinor connection (4.20) is not directly applicable in well-based spaces. This situation is easily remedied, however. It was shown in Chapter I that (4.24) determines the affine connection $L^\alpha{}_{\beta\gamma}$ uniquely, and in fact that we have

$$L^\alpha{}_{\beta\gamma} = \Gamma^\alpha{}_{\beta\gamma}(h_{\eta\pi}) + T^\alpha{}_{\beta\gamma}, \tag{4.26}$$

where

$$T^\alpha{}_{\beta\gamma} = \tfrac{1}{2}\, h^{\alpha\sigma}\, (Q_{\beta\sigma\gamma} - Q_{\sigma\gamma\beta} + Q_{\gamma\beta\sigma}). \tag{4.27}$$

Now, we have

$$\mathfrak{G}_{\alpha;\beta} = \nabla_\beta \mathfrak{G}_\alpha - T^\rho{}_{\alpha\beta}\, \mathfrak{G}_\rho$$

by definition of the covariant derivative and use of (4.26), and hence

$$\nabla_\beta \mathfrak{G}_\alpha = \mathfrak{G}_{\alpha;\beta} + T^\sigma{}_{\alpha\beta}\, \mathfrak{G}_\sigma. \tag{4.28}$$

Let $\mathfrak{D}_\gamma$ denote the covariant derivative with respect to the following two connections: (a) the connection $L^\alpha{}_{\beta\gamma}$ of a well-based space (4.26); (b) the spinor connection $\mathfrak{A}_\gamma$ associated with such a well-based space. From the definition of the D_γ derivative, we have

$$D_\beta \mathfrak{G}_\alpha = \nabla_\beta \mathfrak{G}_\alpha + \mathfrak{A}_\beta\, \mathfrak{G}_\alpha - \mathfrak{G}_\alpha\, \mathfrak{A}_\beta.$$

Substituting from (4.28) gives

$$D_\beta \mathfrak{G}_\alpha = \mathfrak{G}_{\alpha;\beta} + T^\sigma{}_{\alpha\beta}\, \mathfrak{G}_\sigma + \mathfrak{A}_\beta\, \mathfrak{G}_\alpha - \mathfrak{G}_\alpha\, \mathfrak{A}_\beta,$$

and hence

$$D_\beta \mathfrak{G}_\alpha = \mathfrak{D}_\beta \mathfrak{G}_\alpha + T^\sigma{}_{\alpha\beta}\, \mathfrak{G}_\sigma, \tag{4.29}$$

by the definition of the $\mathfrak{D}_\gamma$ derivative. Thus the generating equation (4.5) of the spinor connection becomes

$$\mathfrak{D}_\beta \mathfrak{G}_\alpha = -T^\sigma{}_{\alpha\beta}\, \mathfrak{G}_\sigma, \qquad (\mathfrak{D}_\beta \mathfrak{N}^\alpha = T^\alpha{}_{\beta\sigma}\, \mathfrak{N}^\sigma). \tag{4.30}$$

From equation (4.30) we see that the spinor connection in a well-based space is directly obtainable from the spinor connection in a metric space through replacing the ∇_β derivative by the semicolon derivative in (4.15) by means of (4.28). This result follows since (4.30) is just a restatement of the generating equation (4.5) in a well-based space. *Thus the relation*

$$4\mathfrak{M}_\beta = \mathfrak{N}^\alpha \mathfrak{G}_{\alpha;\beta} - \mathfrak{G}_{\alpha;\beta} \mathfrak{N}^\alpha + T^\sigma{}_{\alpha\beta} (\mathfrak{N}^\alpha \mathfrak{G}_\sigma - \mathfrak{G}_\sigma \mathfrak{N}^\alpha) \tag{4.31}$$

together with (4.20) *establishes the spinor connection in a well-based space.*

It is to be noted that (4.24) can be recovered by use of this spinor connection. Since

$$h_{\alpha\beta} \mathfrak{J} = 2\mathfrak{G}_{(\alpha} \mathfrak{G}_{\beta)},$$

upon operating on both sides with $\mathfrak{D}_\gamma$ and using (4.30) we obtain

$$\begin{aligned} \mathfrak{D}_\gamma h_{\alpha\beta} \mathfrak{J} &= -2T^\sigma{}_{\alpha\gamma} \mathfrak{G}_{(\sigma} \mathfrak{G}_{\beta)} - 2T^\sigma{}_{\gamma\beta} \mathfrak{G}_{(\sigma} \mathfrak{G}_{\alpha)} \\ &= -(T^\sigma{}_{\alpha\gamma} h_{\sigma\beta} + T^\sigma{}_{\beta\gamma} h_{\alpha\sigma}) \mathfrak{J}. \end{aligned}$$

Now by (4.27) we have

$$-T^\sigma{}_{\alpha\gamma} h_{\sigma\beta} - T^\sigma{}_{\beta\gamma} h_{\alpha\sigma} = -Q_{\gamma\alpha\beta},$$

and hence

$$(\mathfrak{D}_\gamma h_{\alpha\beta} + Q_{\gamma\alpha\beta}) \mathfrak{J} = 0.$$

This, however, is just equation (4.24) since $\mathfrak{D}_\gamma h_{\alpha\beta} = h_{\alpha\beta;\gamma}$ by the definition of the $\mathfrak{D}_\gamma$ derivative.

It is of interest to note the form of $\mathfrak{M}_\beta$ in the field spaces encountered in Chapters VII and VIII. For such field spaces, we have

$$4T^\sigma{}_{\alpha\beta} = f_\alpha \delta^\sigma_\beta + f_\beta \delta^\sigma_\alpha - 3f^\sigma_{.} h_{\alpha\beta}.$$

Substituting this into (4.31) then yields

$$\begin{aligned} 4\mathfrak{M}_\beta = \mathfrak{N}^\alpha \mathfrak{G}_{\alpha;\beta} - \mathfrak{G}_{\alpha;\beta} \mathfrak{N}^\alpha + \tfrac{1}{4} (\mathfrak{F} \mathfrak{G}_\beta - \mathfrak{G}_\beta \mathfrak{F}) \\ + \tfrac{3}{4} h_{\alpha\beta} (\mathfrak{N}^\alpha \mathfrak{F}_{.} - \mathfrak{F}_{.} \mathfrak{N}^\alpha), \end{aligned} \tag{4.32}$$

where the quantities

$$\mathfrak{F} = \mathfrak{N}^\alpha f_\alpha, \qquad \mathfrak{F}_{.} = f^\alpha_{.} \mathfrak{G}_\alpha \tag{4 33}$$

are the spinors associated with f_α and $f^\alpha_{.}$, respectively.

From this point we can proceed in a purely formal manner and define a spin curvature tensor $\mathfrak{R}_{\lambda\mu}$ by

$$\mathfrak{R}_{\lambda\mu} = 2\mathfrak{A}_{[\mu,\lambda]} + 2\mathfrak{A}_{[\lambda}\,\mathfrak{A}_{\mu]} \qquad (4.34)$$

and thereby obtain all results previously derived for curvature tensors in Chapter II.

Appendix B

Hamiltonian Form of the Variant Field Equations

THE STUDY OF systems of partial differential equations is usually significantly simplified if they are written in Hamiltonian† form. Although it is possible to obtain equations in Hamiltonian form by any of several methods, the most direct and useful for our purposes is the method of auxiliary variables introduced by Lanczos.‡

The basic Lagrangian introduced in Chapter III was assumed to have the functional form

$$\mathfrak{L} = \mathfrak{L}(A_{\alpha\beta}, R_{\alpha\beta}, L^{\alpha}{}_{\beta\gamma}, Q_{(\rho)}, Q_{(\rho),\gamma}, \mathbf{x}). \tag{1.1}$$

If we replace $A_{\alpha\beta}$, $R_{\alpha\beta}$, and $Q_{(\rho),\gamma}$ with the variables $a_{\alpha\beta}$, $r_{\alpha\beta}$, and $q_{(\rho)\gamma}$, so that

$$\mathfrak{L} = \underline{\mathfrak{L}}(a_{\alpha\beta}, r_{\alpha\beta}, L^{\alpha}{}_{\beta\gamma}, Q_{(\rho)}, q_{(\rho)\gamma}, x), \tag{1 2}$$

then $a_{\alpha\beta}$, $r_{\alpha\beta}$, $L^{\alpha}{}_{\beta\gamma}$, $Q_{(\rho)}$, and $q_{(\rho)\gamma}$ can be considered as independent, provided the following equations of constraint are satisfied:

$$0 = A_{\alpha\beta} - a_{\alpha\beta} = L^{\sigma}{}_{\sigma\beta,\alpha} - L^{\sigma}{}_{\sigma\alpha,\beta} - a_{\alpha\beta}, \tag{1.3}$$

$$0 = R_{\alpha\beta} - r_{\alpha\beta} = \tfrac{1}{2}\,(L^{\sigma}{}_{\sigma\alpha,\beta} + L^{\sigma}{}_{\sigma\beta,\alpha}) - L^{\sigma}{}_{\alpha\beta,\sigma}$$
$$+ L^{\mu}{}_{\sigma\alpha}\,L^{\sigma}{}_{\mu\beta} - L^{\mu}{}_{\mu\sigma}\,L^{\sigma}{}_{\alpha\beta} - r_{\alpha\beta}, \tag{1.4}$$

$$0 = q_{(\rho)\zeta} - Q_{(\rho),\zeta}. \tag{1.5}$$

† D. G. B. Edelen: "The Invariance Group for Hamiltonian Systems of Partial Differential Equations I. Analysis," *Arch. Rational Mech. Anal.*, **5** (1960), pp. 95–176.

‡ C. Lanczos: "Electricity and General Relativity," *Rev. Mod. Phys.*, **29** (1957), pp. 337–350.

Multiplying the constraint equations by the Lagrangian undetermined multipliers $H^{\alpha\beta}$, $I^{\alpha\beta}$, and $W^{(\rho)\zeta}$, respectively, we obtain the free variational problem

$$\delta \int_{D^*} \mathfrak{L}^* \, dV = 0 \tag{1.6}$$

in the variables $a_{\alpha\beta}$, $H^{\alpha\beta}$, $r_{\alpha\beta}$, $I^{\alpha\beta}$, $L^{\alpha}{}_{\beta\gamma}$, $Q_{(\rho)}$, $q_{(\rho)\zeta}$, and $W^{(\rho)\zeta}$, where

$$\begin{aligned} \mathfrak{L}^* &= (L^{\sigma}{}_{\sigma\beta,\alpha} - L^{\sigma}{}_{\sigma\alpha,\beta})\, H^{\alpha\beta} \\ &\quad + [\tfrac{1}{2}\,(L^{\sigma}{}_{\sigma\alpha,\beta} + L^{\sigma}{}_{\sigma\beta,\alpha}) - L^{\sigma}{}_{\alpha\beta,\sigma}]\, I^{\alpha\beta} \\ &\quad + Q_{(\rho),\zeta}\, W^{(\rho)\zeta} - \tilde{H}, \end{aligned} \tag{1.7}$$

and where

$$\begin{aligned} \tilde{H} &= a_{\alpha\beta}\, H^{\alpha\beta} + (L^{\mu}{}_{\mu\sigma}\, L^{\sigma}{}_{\alpha\beta} - L^{\mu}{}_{\sigma\beta}\, L^{\sigma}{}_{\mu\alpha} + r_{\alpha\beta})\, I^{\alpha\beta} \\ &\quad + q_{(\rho)\zeta}\, W^{(\rho)\zeta} - \underline{\mathfrak{L}}(a_{\alpha\beta}, r_{\alpha\beta}, L^{\alpha}{}_{\beta\gamma}, Q_{(\rho)}, Q_{(\rho),\zeta}, \boldsymbol{x}). \end{aligned} \tag{1.8}$$

Evaluating the variation of

$$\int_{D^*} \mathfrak{L}^* \, dV$$

with respect to $a_{\alpha\beta}$, $r_{\alpha\beta}$, and $q_{(\rho)\zeta}$, respectively, and equating the results to zero in accordance with (1.6) yields

$$H^{\alpha\beta} = \underline{\mathfrak{L}}_{,a_{\alpha\beta}}, \quad I^{\alpha\beta} = \underline{\mathfrak{L}}_{,r_{\alpha\beta}}, \quad W^{(\rho)\zeta} = \underline{\mathfrak{L}}_{,q_{(\rho)\zeta}}. \tag{1.9}$$

Comparing (1.9) with (III-2.10) and noting that

$$\mathfrak{L}_{,A_{\alpha\beta}} = \underline{\mathfrak{L}}_{,a_{\alpha\beta}}, \quad \mathfrak{L}_{,R_{\alpha\beta}} = \underline{\mathfrak{L}}_{,r_{\alpha\beta}},$$

we see that the quantities $H^{\alpha\beta}$, $I^{\alpha\beta}$, and $W^{(\rho)\zeta}$ introduced in this appendix as Lagrangian undetermined multipliers are the same as the field variables $H^{\alpha\beta}$, $I^{\alpha\beta}$, and $W^{(\rho)\zeta}$ introduced in Chapter III. If

$$\det(\underline{\mathfrak{L}}_{,a_{\alpha\beta},a_{\eta\mu}}) \neq 0,$$

$$\det(\underline{\mathfrak{L}}_{,r_{\alpha\beta},r_{\eta\mu}}) \neq 0,$$

$$\det(\underline{\mathfrak{L}}_{,q_{(\rho)\zeta},q_{(\varphi)\psi}}) \neq 0,$$

then (1.9) can be solved for $a_{\alpha\beta}$, $r_{\alpha\beta}$, and $q_{(\rho)\zeta}$ as functions of $H^{\alpha\beta}$, $I^{\alpha\beta}$, $L^{\alpha}{}_{\beta\gamma}$, $W^{(\rho)\zeta}$, and $Q_{(\rho)}$. Using these solutions, we can eliminate $a_{\alpha\beta}$, $r_{\alpha\beta}$, and $q_{(\rho)\zeta}$ in (1.8), thus obtaining

$$\tilde{H} = \tilde{H}(H^{\alpha\beta}, I^{\alpha\beta}, L^{\alpha}{}_{\beta\gamma}, W^{(\rho)\zeta}, Q_{(\rho)}, \boldsymbol{x}). \tag{1.10}$$

Evaluating the variation of

$$\int_{D^*} \mathfrak{L}^* \, dV$$

with respect to $H^{\alpha\beta}$, $I^{\alpha\beta}$, $L^{\alpha}{}_{\beta\gamma}$, $W^{(\rho)\zeta}$, $Q_{(\rho)}$, using the symmetry of $L^{\alpha}{}_{\beta\gamma}$, $I^{\alpha\beta}$, $\delta L^{\alpha}{}_{\beta\zeta}$, and $H^{\alpha\beta}$, and equating the results to zero in accordance with equation (1.6) gives

$$0 = L^{\sigma}{}_{\sigma\beta,\alpha} - L^{\sigma}{}_{\sigma\alpha,\beta} - \tilde{H}_{,H^{\alpha\beta}}, \tag{1.11}$$

$$0 = \tfrac{1}{2}\,(L^{\sigma}{}_{\sigma\alpha,\beta} + L^{\sigma}{}_{\sigma\beta,\alpha}) - L^{\sigma}{}_{\alpha\beta,\sigma} - \tilde{H}_{,I^{\alpha\beta}}, \tag{1.12}$$

$$0 = Q_{(\rho),\zeta} - \tilde{H}_{,W^{(\rho)\zeta}}, \tag{1.13}$$

$$0 = -W^{(\rho)\zeta}{}_{,\zeta} - \tilde{H}_{,Q_{(\rho)}}, \tag{1.14}$$

$$\begin{aligned} 0 = {} & -2\tilde{H}_{,L^{\zeta}{}_{\eta\lambda}} + 2\delta^{\lambda}_{\zeta} H^{\eta\sigma}{}_{,\sigma} + 2\delta^{\eta}_{\zeta} H^{\lambda\sigma}{}_{,\sigma} \\ & - \delta^{\lambda}_{\zeta} I^{\eta\sigma}{}_{,\sigma} - \delta^{\eta}_{\zeta} I^{\lambda\sigma}{}_{,\sigma} + 2I^{\eta\lambda}{}_{,\zeta}. \end{aligned} \tag{1.15}$$

Equations (1.11)–(1.15) *are the Hamiltonian form of the variant field equations.*

Contracting equation (1.15) gives

$$0 = 2\tilde{H}_{,L^{\zeta}{}_{\zeta\eta}} + 10H^{\eta\zeta}{}_{,\zeta} - 3I^{\eta\zeta}{}_{,\zeta},$$

from which we obtain

$$H^{\eta\zeta}{}_{,\zeta} = \tfrac{1}{10}\,(3I^{\eta\zeta}{}_{,\zeta} + 2\tilde{H}_{,L^{\zeta}{}_{\zeta\eta}}). \tag{1.16}$$

Thus, the Hamiltonian form of the Maxwellian current is given by

$$S^{\eta} = \tfrac{1}{10}\,(3I^{\eta\zeta}{}_{,\zeta} + \tilde{H}_{,L^{\zeta}{}_{\zeta\eta}}). \tag{1.17}$$

Eliminating $H^{\eta\zeta}{}_{,\zeta}$ between (1.16) and (1.15) gives

$$\begin{aligned} & I^{\zeta\eta}{}_{,\mu} - \tfrac{1}{5}\,\delta^{\eta}_{\mu} I^{\zeta\sigma}{}_{,\sigma} - \tfrac{1}{5}\,\delta^{\zeta}_{\mu} I^{\eta\sigma}{}_{,\sigma} \\ & \quad = (\tfrac{1}{5}\,\delta^{\zeta}_{\mu}\,\tilde{H}_{,L^{\sigma}{}_{\sigma\eta}} + \tfrac{1}{5}\,\delta^{\eta}_{\mu}\,\tilde{H}_{,L^{\sigma}{}_{\sigma\zeta}} - \tilde{H}_{,L^{\mu}{}_{\zeta\eta}}). \end{aligned} \tag{1.18}$$

Equations (1.16) and (1.18) are equivalent to (III-1.7) and (III-1.8) and are considerably simpler to solve, since $\tilde{H}_{,L^{\mu}{}_{\zeta\eta}}$ will, in general, be independent of $H^{\alpha\beta}$.

BIBLIOGRAPHY

Bibliography

THIS BIBLIOGRAPHY is intended as a chronological survey of field theory and related topics. The entries have been selected to cover as many of the proposed theories as possible without going too far astray. No completeness is claimed, and it must be admitted that references on electromagnetic phenomena predominate. For a more complete bibliography on general relativity, see M. Lecat: *Bibliographie de la relativité*, Lamertin, Brussels (1924), and J. L. Synge: *Relativity: The General Theory*, North-Holland Publishing Co., Amsterdam (1960). Unfortunately, there is no similar bibliographic source on quantum field theory or general field-theoretic descriptions of natural phenomena. It is hoped that the references listed here cover the various avenues that have been pursued during the past forty years and provide a perspective for viewing current work.

1916

Einstein, A.: "Die Grundlage der allgemeinen Relativitätstheorie," *Ann. Physik*, **49**, 769.

Einstein, A.: "Eine neue Deutig der Maxwellschen Feldgleichungen der Elektrodynamik," *Preuss. Akad. Wiss. Sitz.*, 184.

Einstein, A.: "Hamiltonsches Prinzip und allgemeine Relativitätstheorie," *Preuss. Akad. Wiss. Sitz.*, 1111.

1917

Einstein, A.: "Kosmologische Betrachtungen zur allgemeinen Relativitätstheorie," *Preuss. Akad. Wiss. Sitz.*, 142.

1918

Palatini, A.: "Deduzione invariantiva della equazioni gravitazionali dal principio di Hamilton," *Rend. Circ. Mat. Palermo*, **43**, 203.

Weyl, H.: "Gravitation und Elektrizität," *Preuss. Akad. Wiss. Sitz.*, 465.

1919

Einstein, A.: "Spielen Gravitationsfelder im Ausser der materiellen Elementarteilchen eine wesentliche Rolle?" *Preuss. Akad. Wiss. Sitz.*, 349.

Pauli, W.: "Zur Theorie der Gravitation und Elektrizität von H. Weyl," *Physik. Zs.*, **20**, 457.

1920

Buhl, A.: "Sur les symétries du champ électromagnétique et gravifique," *C. R. Acad. Sci. Paris*, **171**, 345.

Donder, T. de: *Théorie du champ électromagnétique de Maxwell et du champ gravifique d'Einstein*, Gauthier-Villars et Cie, Paris.

Weyl, H.: "Elektrizität und Gravitation," *Phys. Z.*, **21**, 649.

1921

Eddington, A. S.: "A Generalization of Weyl's Theory of the Electromagnetic and Gravitational Fields," *Proc. Roy. Soc. London*, **A99**, 104.

Weyl, H.: "Feld und Materie," *Ann. Physik*, **65**, 541.

1922

Becquerel, J.: *La principe de la relativité et la théorie de la gravitation*, Gauthier-Villars et Cie, Paris.

Friedman, A.: "Über die Krümmung des Raumes," *Z. Physik*, **10**, 377.

Kottler, F.: "Maxwellsche Gleichungen und Metrik," *Wien. Ber.*, **131**, 119.

Weyl, H.: *Space-Time-Matter*, Methuen & Co., Ltd., London.

1924

Donder, T. de: *La gravifique de Weyl-Eddington-Einstein*, Gauthier-Villars et Cie, Paris.

Eddington, A. S.: *The Mathematical Theory of Relativity*, 2d ed., Cambridge University Press, London.

Lecat, M.: *Bibliographie de la relativité*, Lamertin, Brussels.

Rainich, G. Y.: "Electrodynamics in General Relativity Theory," *Proc. Nat. Acad. Sci. USA*, **10**, 124, 294.

1925

Cartan, E.: *La géométrie des espaces de Riemann*, Gauthier-Villars et Cie, Paris.

Rainich, G. Y.: "Electrodynamics in General Relativity," *Trans. Amer. Math. Soc.*, **27**, 106.

1927

Beck, G.: "La propagation des ondes électromagnétiques dans la théorie de la relativité générale," *Arch. Sci. Phys. Mat. Genève*, **8**, 75.

Birkhoff, G. D.: "A Theory of Matter and Electricity," *Proc. Nat. Acad. Sci. USA*, **13**, 160.

1928

Whittaker, E. T.: "The Influence of Gravitation on Electromagnetic Phenomena," *J. London Math. Soc.*, **3**, 137.

1929

Fock, V. A.: "Sur les équations de Dirac dans la théorie de relativité générale," *C. R. Acad. Sci. Paris*, **189**, 25.

Fokker, A. D.: *Relativiteitstheorie*, P. Noordhoff, Ltd., Groningen, Holland.

Weyl, H.: "Gravitation and the Electron," *Proc. Nat. Acad. Sci. USA*, **15**, 323.

1930

Thomas, T. Y.: "On the Unified Field Theory," *Proc. Nat. Acad. Sci. USA*, **16**, 761.

1931

Alexandrow, W.: "Über die allgemein koordinateninvarianten Gleichungen der Wellenmechanik (Materie und Gravitation)," *Z. Physik*, **68**, 813.

Eddington, A. S.: "On the Value of the Cosmological Constant," *Proc. Roy. Soc. London*, **A133**, 605.

Kunii, S.: "On a Unified Theory of Gravitational and Electromagnetic Fields," *Mem. Coll. Sci. Kyoto Univ.*, **A14**, 195.

Lanczos, C.: "Elektromagnetismus als natürliche Eigenschaft der Riemannschen Geometrie," *Z. Physik*, **73**, 147.

Weyl, H.: "Geometrie und Physik," *Naturwissenschaften*, **19**, 49.

1932

Donder, T. de, and Y. Dupont: "Théorie relativiste de l'élasticité et de l'électromagnétostriction," *Acad. Roy. Belg. Bull. Cl. Sci.*, **18**, 680, 782, 899.

Lanczos, C.: "Electricity as a Natural Property of Riemannian Geometry," *Phys. Rev.*, **39**, 188.

Schouten, J. A., and D. van Dantzig: "Zur generellen Feldtheorie. Diracsche Gleichungen und Hamiltonsche Funktion," *Akad. Wetensch. Amsterdam Proc.*, **35**, 843.

Viney, I. E., and G. H. Livens: "Gravitation and Electricity," *Phil. Mag.*, **14**, 243.

1933

Eddington, A. S.: *The Expanding Universe*, Cambridge University Press, London.

Etherington, I. M. H.: "On the Definition of Distance in General Relativity," *Phil. Mag.*, **15**, 761.

Gormley, P. G.: "On Staneo's Unified Theory of Gravitation and Electricity," *Proc. Edinburgh Math. Soc.*, **3**, 269.

Robertson, H. P.: "Relativistic Cosmology," *Rev. Mod. Phys.*, **5**, 62.

Schouten, J. A.: "Zur generellen Feldtheorie; Ableitung des Impulsenergiestromprojektors aus einem Variationsprinzip," *Z. Physik*, **81**, 129.

Soh, H. P.: "A Theory of Gravitation and Electromagnetism," *J. Math. Phys.*, **12**, 298.

Takeuchi, T.: "Universe without Curvature," *Proc. Phys.-Math. Soc. Japan*, **15**, 217.

Veblen, S.: *Projective Relativitätstheorie*, Springer-Verlag, Berlin.

1934

Born, M., and L. Infeld: "Foundations of the New Field Theory," *Proc. Roy. Soc. London*, **A144**, 425.

Dantzig, D. van: "Electromagnetism Independent of Metrical Geometry. I. The Foundations. II. Variational Principles and Further Generalizations of the Theory. III. Mass and Momentum. IV. Momentum and Energy; Waves," *Akad. Wetensch. Amsterdam*, **37**, 521, 526, 643, 825.

Dantzig, D. van: "The Fundamental Equations of Electromagnetism, Independent of Metric Geometry," *Proc. Cambridge Philos. Soc.*, **30**, 421.

Meksyn, D.: "A Unified Field Theory. I. Electromagnetic Field. II. Gravitation," *Phil. Mag.*, **17**, 99, 476.

Thomas, T. Y.: *The Differential Invariants of Generalized Spaces*, Cambridge University Press, London.

Tolman, R. C.: *Relativity, Thermodynamics, and Cosmology*, Clarendon Press, Oxford.

1935

Donder, T. de: *Théorie invariantive du calculus des variations*, Gauthier-Villars et Cie, Paris.

Flint, H. T.: "A Relativistic Basis for the Quantum Theory III," *Proc Roy. Soc. London*, **A150**, 421.

Halpern, O., and G. Heller: "On the Dirac Electron in a Gravitational Field," *Phys. Rev.*, **48**, 434.

Hosokawa, T.: "On the Foundation of the Geometry in Microscopic and Macroscopic Space," *J. Sci. Hiroshima Univ.*, **A5**, 141.

Schouten, J. A.: "Zur generellen Feldtheorie. Raumzeit und Spinraum," (*G.F.V.*) *Z. Physik*, **81**, 405.

1936

Bronstein, M.: "Quantization of Gravitational Waves," *Z. Eksper. Teoret. Fis.*, **6**, 195.

Eddington, A. S.: "The Cosmical Constant and the Recession of the Nebulae," *Amer. J. Math.*, **59**, 1.

Eddington, A. S.: *The Relativity Theory of Protons and Electrons*, Cambridge University Press, London.

Mimura, Y.: "Geometrization of the Laws of Physics," *J. Sci. Hiroshima Univ.*, **A7**, 81.

Page, L.: "A New Relativity. I. Fundamental Principles and Transformation between Accelerated Systems," *Phys. Rev.*, **49**, 254.

Proca, A.: "Sur la théorie ondulatoire des électrons positifs et négatifs," *J. Phys. Radium*, **7**, 347.

Robb, A. A.: *Geometry of Space and Time*, Cambridge University Press, London.

Robertson, H. P.: "An Interpretation of Page's 'New Relativity,' " *Phys. Rev.*, **49**, 755.

1937

Mimura, Y., and T. Hosokawa: "Physics and Geometry," *J. Sci. Hiroshima Univ.*, **A7**, 249.

1938

Benedictus, W.: "Les équations de Dirac dans un espace à métrique Riemannienne," *C. R. Acad. Sci. Paris*, **206**, 1951.

Cartan, E.: *Leçons sur la théorie des spineurs*, *I*, *II*, Hermann et Cie, Paris.

Einstein, A., L. Infeld, and B. Hoffman: "The Gravitational Equations and the Problem of Motion," *Ann. of Math.*, **39**, 65.

Weiss, P.: "On the Hamilton-Jacobi Theory and Quantization of a Dynamical Continuum," *Proc. Roy. Soc. London*, **A169**, 102.

1939

Lees, A.: "The Electron in General Relativity Theory," *Phil. Mag.*, **28**, 385.

Lichnerowicz, A.: *Problèmes globaux en mécanique relativiste*, Hermann et Cie, Paris.

Mimura, Y., and T. Hosokawa: "Space, Time, and the Laws of Nature," *J. Sci. Hiroshima Univ.*, **A9**, 217.

1940

Copson, E. T.: "On Electrostatics in a Gravitational Field," *Proc. Roy. Soc. London*, **A118**, 184.

Infeld, L., and P. R. Wallace: "The Equations of Motion in Electrodynamics," *Phys. Rev.*, **57**, 797.

Schrodinger, E.: "Maxwell's and Dirac's Equations in the Expanding Universe," *Proc. Roy. Irish Acad.*, **A46**, 25.

1941

Wallace, P. R.: "Relativistic Equations of Motion in Electromagnetic Theory," *Amer. J. Math.*, **63**, 729.

1942

Bergmann, P. G.: *Introduction to the Theory of Relativity*, Prentice-Hall, Inc., Englewood Cliffs, N.J.

Lanczos, C.: "Matter Waves and Electricity," *Phys. Rev.*, **61**, 713.

Martin, D.: "On the Method of Extending Dirac's Equation to General Relativity," *Proc. Edinburgh Math. Soc.*, **7**, 39.

1943

Birkhoff, G. D.: "Matter, Electricity and Gravitation in Flat Spacetime," *Proc. Nat. Acad. Sci. USA*, **29**, 231.

Eddington, A. S.: "The Combination of Relativity Theory and Quantum Theory," *Comm. Dublin Inst. Adv. Studies*, **A2**.

Lichnerowicz, A.: "Sur les équations relativistes de l'électromagnétisme," *Ann. Sci. Ecole Norm. Sup.*, **60**, 247.

1944

Schild, A.: "On Milne's Theory of Gravitation," *Phys. Rev.*, **66**, 340.

Weyl, H.: "Comparison of a Degenerate Form of Einstein's with Birkhoff's Theory of Gravitation," *Proc. Nat. Acad. Sci. USA*, **30**, 205.

1945

Brekhovskich, L. M.: "Radiation of Gravitational Waves by Electromagnetic Waves," *C. R. (Dokl.) Acad. Sci. USSR*, **N.S.49**, 482.

Einstein, A.: "A Generalization of the Relativistic Theory of Gravitation," *Ann. of Math.*, **46**, 578.

Jaiswal, J. P.: "On the Electric Potential of a Single Electron in a Gravitational Field," *Proc. Benares Math. Soc.*, **N.S.7**, 17.

Mariani, J.: "Electromagnétisme et relativité. Le magnétisme terrestre comme conséquence de la relativité générale," *Cahiers de Physique*, **28**, 23.

1946

Dittrich, A.: "Die Hamilton-Jacobische Methode in der Einsteinschen Mechanik," *Casopis*, **53**, 38.

1947

Hoffman, B.: "The Vector Meson Field and Projective Relativity," *Phys. Rev.*, **72**, 458.

1948

Bondi, H., and T. Gold: "The Steady-state Theory of the Expanding Universe," *Roy. Astr. Soc. (London), M. N.*, **108**, 253.

Buchdahl, H. A.: "On Eddington's Higher Order Equations of the Gravitational Field," *Proc. Edinburgh Math. Soc.*, **8**, 89.

Einstein, A.: "A Generalized Theory of Gravitation," *Rev. Mod. Phys.*, **20**, 35.

Hoffman, B.: "The Gravitational, Electromagnetic and Vector Meson Fields and the Similarity Geometry," *Phys. Rev.*, **73**, 30.

Jordan, P.: "Fünfdimensionale Kosmologie," *Astr. Nachr.*, **276**, 193.

Jordan, P.: *Projective Relativitätstheorie und Kosmologie*, Dietrichische Verlagsbuchhandlund, Wiesbaden.

Walker, A. G.: "Foundations of Relativity, I, II," *Proc. Roy. Soc. Edinburgh*, **A62**, 319.

1949

Alpher, R. A., and R. C. Herman: "Remarks on the Evolution of the Expanding Universe," *Phys. Rev.*, **75**, 1089.

Bergmann, P. G., and J. H. M. Brunings: "Non-linear Field Theories II. Canonical Equations and Quantization," *Rev. Mod. Phys.*, **21**, 480.

Clark, G. L.: "The Mechanics of Continuous Matter in the Relativity Theory," *Proc. Roy. Soc. Edinburgh*, **A62**, 434.

Finzi, B., and M. Pastori: *Calculo tensoriale e applicazioni*, Zanichelli, Bologna.

Gel'man, E. E.: "Real Spinors in the General Theory of Relativity," *Leningrad Gos. Univ. Ucenye Zapiski 120, Ser. Fiz. Nauk*, **7**, 79.

McVittie, G. C.: *Cosmological Theory*, 2d ed., Methuen & Co., Ltd., London.

Sato, I.: "An Attempt to Unite the Quantum of Wave Field with the Theory of General Relativity," *Sci. Rep. Tôhoku Univ.*, **I33**, 30.

Schouten, J. A.: "On Meson Fields and Conformal Transformations," *Rev. Mod. Phys.*, **21**, 421.

1950

Bergmann, D. G., R. Penfield, R. Schiller, and H. Zatzkis: "The Hamiltonian of the General Theory of Relativity with Electromagnetic Field," *Phys. Rev.*, **80**, 81.

Pirani, F. A. E., and A. Schild: "On the Quantization of Einstein's Gravitational Field Equations," *Phys. Rev.*, **79**, 986.

Rainick, G. Y.: *Mathematics of Relativity*, John Wiley & Sons, New York.

Schrodinger, E.: *Space-Time Structure*, Cambridge University Press, London.

Synge, J. L.: "Electromagnetism without Metric," *Proc. Symposia on Appl. Math., Amer. Math. Soc.*, **2**, 21.

1951

Debever, R.: "Les espaces de l'électromagnétisme," *Colloque de géometrie différentielle*, Tenu à Louvain, 217.

García, G.: "Los fundamentos y construcción de una teoría de la relatividad general. El concepto de tiempo. La nueva lay completa de la gravitación universal. Las ecuaciones diferenciales del movimiento de la nueva dinâmica," *Actas de la Acad. Nat. de Ciencias Exactas, Fís. y Nat. Lima*, **14**, 3.

Graef Fernandez, C.: "Birkhoff's Theory of Gravitation," *Symposium sobre algunos problemas mathemáticos que se están estudiano en Latino America*, **121**.

Landau, L., and E. Lifshitz: *The Classical Theory of Fields*, Addison-Wesley Publishing Company, Inc., Cambridge, Mass.

McCrea, W. H.: "Relativity Theory and the Creation of Matter," *Proc. Roy. Soc. London*, **A206**, 562.

Synge, J. L.: "The Relativity Theory of A. N. Whitehead," *Inst. Fluid Mech. and Appl. Math., Univ. Maryland Lecture Ser.*, **5**.

Tonnelat, M. A.: "Théorie unitaire affine du champ physique," *J. Phys. Radium*, **12**, 81.

1952

Bondi, H.: *Cosmology*, Cambridge University Press, London.

Gupta, S. N.: "Quantization of Einstein's Gravitational Field: Linear Approximation," *Proc. Phys. Soc.*, **A65**, 161.

Kursunoğlu, B.: "Gravitation and Electrodynamics," *Phys. Rev.*, **88**, 1369.

Møller, C.: *The Theory of Relativity*, Oxford University Press, New York.

Pirani, F. A. E., A. Schild, and R. Skinner: "Quantization of Einstein's Gravitational Field Equations. II," *Phys. Rev.*, **87**, 452.

1953

Bergmann, P. G., and R. Thomson: "Spin and Angular Momentum in General Relativity," *Phys. Rev.*, **89**, 400.

Einstein, A.: "A Comment on a Criticism of Unified Field Theory," *Phys. Rev.*, **89**, 321.

Goldberg, J. N.: "Strong Conservation Laws and Equations of Motion in Covariant Field Theories," *Phys. Rev.*, **89**, 263.

Hoffman, B.: "The Similarity Theory of Relativity and the Dirac-Schrodinger Theory of Electrons. I, II," *Phys. Rev.*, **89**, 52.

Infeld, L., and J. Plebanski: "Electrodynamics without Potentials," *Acta Phys. Polon*, **12**, 123.

Ingrahm, R. L.: "Spinor Relativity," *Il Nuovo Cimento*, **10**, 1.

Kirkwood, R. L.: "The Physical Basis of Gravitation," *Phys. Rev.*, **92**, 1557.

Nariai, H.: "Some Remarks on Jordan's Projective Relativity," *Sci. Rep. Tôhoku Univ.*, **I37**, 423.

Riesz, M.: "L'équation de Dirac en relativité générale," *Tolfte Skandinaviska Matematikerkongressen, Lund*, 241.

Stephenson, G.: "Dirac's Electrodynamics and Einstein's Unified Field Theory," *Il Nuovo Cimento*, **10**, 1595.

Stephenson, G., and C. W. Kilmister: "A Unified Theory of Gravitation and Electromagnetism," *Il Nuovo Cimento*, **10**, 230.

Takasu, T.: "A Necessary Unitary Field Theory as a Non-holonomic Lie Geometry Realized in Three-dimensional Cartesian Space and Its Quantum Mechanics," *Yokohama Math. J.*, **1**, 263.

Taylor, N. W.: "The Relativistic Electromagnetic Equations in a Material Medium," *Australian J. Phys.*, **6**, 1.

1954

Chase, D. M.: "Equations of Motion of a Charged Test Particle in General Relativity," *Phys. Rev.*, **95**, 243.

Duan', I-Si.: "Generalizations of the Regular Solutions of Einstein's

Equations of Gravity and of Maxwell's of Electromagnetism for a Point Charge," *Ž. Eksper. Teoret. Fiz.*, **27**, 756.

Good, R. H., Jr.: "Hamiltonian Mechanics of Fields," *Phys. Rev.*, **93**, 239.

Gupta, S. N.: "Gravitation and Electromagnetism," *Phys. Rev.*, **96**, 1683.

Hatalkar, M. H.: "Theory of Elementary Particles in General Relativity," *Phys. Rev.*, **94**, 1472.

Kilmister, C. W., and G. Stephenson: "An Axiomatic Criticism of Unified Field Theory," *Il Nuovo Cimento*, **11** (supp), 91, 118.

Kirkwood, R. L.: "Gravitational Field Equations," *Phys. Rev.*, **95**, 1051.

Popovici, A.: "Déduction variationnelle des équations gravifiques et électromagnétiques, conformes covariantes de IIe ordre," *Acad. R. P. Romîne Bul. Sti. Sect. Sti. Mat. Fiz.*, **6**, 65.

Raynor, C. B.: "The Application of the Whitehead Theory of Relativity of Non-static, Spherically Symmetric Systems," *Proc. Roy. Soc. London*, **A222**, 509.

Taub, A. H.: "General Relativistic Variational Principle for Perfect Fluids," *Phys. Rev.*, **94**, 1468.

1955

Arzeliès, H.: *La cinématique relativiste*, Gauthier-Villars et Cie, Paris.

Belinfante, F. J.: "Attempts of Quantization of the Gravitational Field," *Rev. Mexicana Fis.*, **4**, 192.

Bertotti, B.: "On the Motion of Charged Particles in General Relativity," *Il Nuovo Cimento*, **2**, 231.

Busemann, H.: *The Geometry of Geodesics*, Academic Press, Inc., New York.

Einstein, A.: *The Meaning of Relativity*, 5th ed., Princeton University Press, Princeton, N. J.

Hennequin, F.: "Interprétation de la théorie de Y. Thiry dans une métnique conforme," *C. R. Acad. Sci. Paris*, **240**, 2378.

Infeld, L.: "The History of Relativity Theory," *Rend. Mat. e Appl.*, **13**, 270.

Kustaanheimo, P.: "Some Remarks on the General Relativity Theory of Birkhoff," *Soc. Sci. Fenn. Comment. Phys.-Math.*, **17**, nr. 11.

Lichnerowicz, A.: *Théories relativistes de la gravitation et de l'électromagnétisme*, Masson, Paris.

Rayner, C. B.: "Whitehead Theory of Gravitation," *Proc. Roy. Soc. London*, **A232**, 135.

Renandie, J.: "Théorie unitaire à six dimensions. Equations du champ. Interprétation pour la champ mésonique-électromagnétique," *C. R. Acad. Sci. Paris*, **240**, 399, 2380.

Robertson, H. P.: "The Theoretical Aspects of the Nebular Redshift," *Publ. Astr. Soc. Pacific*, **67**, 82.

Schweber, S. S., H. A. Bethe, and F. de Hoffman: *Mesons and Fields. Volume I, Fields*, Row, Peterson and Co., White Plains, New York.

Szekeres, G.: "New Formulation of the General Theory of Relativity," *Phys. Rev.*, **97**, 212.

Urbah, V. Y.: "Generalized Theory of Vector Fields," *Dokl. Akad. Nauk SSSR*, **101**, 1043.

1956

Bertotti, B.: "Gravitational Motion and Hamilton's Principle," *Il Nuovo Cimento*, **3**, 655.

Debever, R.: "Etude géométrique du tenseur de Riemann-Christoffel des espaces de Riemann à quatre dimensions," *Acad. Roy. Belg. Bull. Cl. Sci.*, **42**, 313, 608.

Eisenhart, L. P.: "A Unified Theory of General Relativity of Gravitation and Electromagnetism," *Proc. Nat. Acad. Sci. USA*, **42**, 249, 646, 878.

Flint, H. T., and E. M. Williamson: "A Relativistic Theory of Charged Particles in an Electromagnetic and Gravitational Field," *Il Nuovo Cimento*, **3**, 551.

Hoyl, F., and A. Sandage: "The Second-order Term in the Redshift-magnitude Relation," *Publ. Astr. Soc. Pacific*, **68**, 301.

Maurer-Tison, F.: "Théorie unitaire et électromagnétisme dans la matière," *C. R. Acad. Sci. Paris*, **242**, 3042.

McVittie, G. C.: *General Relativity and Cosmology*, John Wiley & Sons, New York.

Schild, A.: "On Gravitational Theories of the Whitehead Type," *Proc. Roy. Soc. London*, **A235**, 202.

Schrödinger, E.: *Expanding Universes*, Cambridge University Press, London.

Umazawa, H.: *Quantum Field Theory*, North-Holland Publishing Co., Amsterdam.

1957

Arnowitt, R. L.: "Phenomenological Approach to a Unified Field Theory," *Phys. Rev.*, **105**, 735.

Bargmann, V.: "Relativity," *Rev. Mod. Phys.*, **29**, 161.

Belinfante, F. J., D. I. Caplan, and W. L. Kennedy: "Quantization of the Interacting Fields of Electrons, Electromagnetism and Gravity," *Rev. Mod. Phys.*, **29**, 518.

Brahmachary, R. L.: "A Class of Exact Solutions of the Combined Gravitational and Electro-magnetic Field Equations of General Relativity," *Il Nuovo Cimento*, **6**, 1502.

Brill, D. R., and J. A. Wheeling: "Interaction of Neutrinos and Gravitational Fields," *Rev. Mod. Phys.*, **29**, 465.

Conference on the Role of Gravitation in Physics, WADC TR 57-216, ASTIA Doc. No. AD 118180.

Deser, S.: "General Relativity and the Divergence Problem in Quantum Field Theory," *Rev. Mod. Phys.*, **29**, 417.

Dicke, R. H.: "Gravitation without a Principle of Equivalence," *Rev. Mod. Phys.*, **29**, 363.

Gupta, S. N.: "Einstein's and Other Theories of Gravitation," *Rev. Mod. Phys.*, **29**, 334.

Hlavatý, V.: *Geometry of Einstein's Unified Field Theory*, P. Noordhoff, Ltd., Groningen, Holland.

Kursunoğlu, B.: "Correspondence in the Generalized Theory of Gravitation," *Rev. Mod. Phys.*, **29**, 412.

Kustaanheimo, P.: "On the Use of a Gravitational Vector Potential in the Relativity Theory of Birkhoff," *Ann. Acad. Sci. Fenn.*, **A228**.

Lanczos, C.: "Electricity and General Relativity," *Rev. Mod. Phys.*, **29**, 337.

Misner, C. W.: "Feynman Quantization of General Relativity," *Rev. Mod. Phys.*, **29**, 497.

Misner, C. W., and J. A. Wheeler: "Classical Physics as Geometry," *Ann. Physics*, **2**, 525.

Taub, A. H.: "Singular Hypersurfaces in General Relativity," *Ill. J. Math.*, **1**, 370.

Wigner, E. P.: "Relativistic Invariance and Quantum Phenomena," *Rev. Mod. Phys.*, **29**, 255.

1958

Callaway, J.: "Klein-Gorden and Dirac Equations in General Relativity," *Phys. Rev.*, **112**, 290.

Davidson, W.: "The Red Shift-Magnitude Relation in Observational Cosmology," *Roy. Astr. Soc.* (*London*), *M. N.*, **119**, 54.

Dirac, P. A. M.: "The Theory of Gravitation in Hamiltonian Form," *Proc. Roy. Soc. London*, **A246**, 333.

Fletcher, J. G.: "Dirac Matrices in Riemannian Space," *Il Nuovo Cimento*, **8**, 451.

Flint, H. T., and E. M. Williamson: "The Theory of Relativity, the Electromagnetic Theory and the Quantum Theory," *Il Nuovo Cimento*, **8**, 680.

Goldberg, J. N.: "Conservation Laws in General Relativity," *Phys. Rev.*, **111**, 315.

Kustaanheimo, P.: "On a Unified Field Theory Based on the Special Theory of Relativity," *Soc. Sci. Fenn. Comment. Phys.-Math.*, **21**, nr. 4.

Kustaanheimo, P.: "Scalar Field Theory as a Theory of Gravitation," *Soc. Sci. Fenn. Comment. Phys.-Math.*, **21**, nr. 3.

Mattig, W.: "Über der Zusammenhang zwischen Rotverschiebund und scheinbarer Helligkeit," *Astr. Nachr.*, **284**, 109.

Peres, A.: "Photons, Gravitons and the Cosmological Constant," *Il Nuovo Cimento*, **8**, 533.

Rzewuski, J.: *Field Theory, Part I, Classical Theory*, Polish Academy of Sciences, Physical Monographs. Panstwowe Wydawnictwo Naukowe, Warszawa.

Sciama, D. W.: "On a Geometrical Theory of the Electromagnetic Field," *Il Nuovo Cimento*, **8**, 417.

Toupin, R. A.: "World Invariant Kinematics," *Arch. Rat. Mech. Anal.*, **1**, 181.

Yilmaz, H.: "New Approach to General Relativity," *Phys. Rev.*, **111**, 1417.

1959

Bertotti, B.: "Uniform Electromagnetic Field in the General Relativity Theory," *Phys. Rev.*, **116**, 1331.

Bogoliubov, N. N., and D. V. Skirkov: *Introduction to the Theory of Quantized Fields*, Interscience Publishers, New York.

Brulin, O., and S. Hjalmars: "The Gravitational Zero Mass Limit of Spin-2 Particles," *Arkiv för Fysik*, **16**, 19.

Cattaneo, C.: "Conservation Laws in General Relativity," *Il Nuovo Cimento*, **13**, 237.

Fock, V. A.: *The Theory of Space, Time and Gravitation*, Pergamon Press, London.

Komar, S.: "Covariant Conservation Laws in General Relativity," *Phys. Rev.*, **113**, 934.

Laurent, B. E.: "A Variational Principle and Conservation Theorems in Connection with the Generally Relativistic Dirac Equation," *Arkiv för Fysik*, **16**, 263.

Laurent, B. E.: "On a Generally Covariant Quantum Theory," *Arkiv för Fysik*, **16**, 237.

Lyttleton, R. A., and H. Bondi: "On the Physical Consequences of a General Excess of Charge," *Proc. Roy. Soc. (London)*, **A252**, 313.

Mercier, A.: *Analytical and Canonical Formalism in Physics*, North-Holland Publishing Co., Amsterdam.

Rosen, G.: "Geometrical Significance of the Einstein-Maxwell Equations," *Phys. Rev.*, **114**, 1179.

Thirring, W.: "Lorentz-invariante Gravitationstheorien," *Fortschritte der Physik*, **7**, 79.

Witten, L.: "Geometry of Gravitation and Electromagnetism," *Phys. Rev.*, **115**, 206.

1960

Brulin, O., and S. Hjalmars: "An Alternative Formulation of the Linearized Classic Theory of Gravitation as a Zero Mass Limit," *Arkiv för Fysik*, **18**, 209.

Edelen, D. G. B.: "The Affine Theory of Electricity and Gravitation," *Arch. Rational Mech. Anal.*, **6**, 1.

Finkelstein, R.: "Spacetime of Elementary Particles," *J. Mathematical Phys.*, **1**, 440.

Fletcher, J. G.: "Local Conservation Laws in Generally Covariant Theories," *Rev. Mod. Phys.*, **32**, 65.

Lichnerowicz, A.: "Ondes et radiations électromagnétiques et gravitationelles en relativité générale," *Ann. Mat. Pura Appl.*, **4**, 1.

Rosen, G.: "Poincaré's Epistemological Sum," *Il Nuovo Cimento*, **16**, 966.

Sakuria, J. J.: "Theory of Strong Interactions," *Ann. of Physics*, **11**, 1.

Scarf, F. L.: "Quantum Electrodynamics in an Expanding Universe," *Il Nuovo Cimento*, **10**, 375.

Synge, J. L.: *Relativity: The General Theory*, North-Holland Publishing Co., Amsterdam.

Truesdell, C., and R. Toupin: "The Classical Field Theories," *Handbuch der Physik*, Band III/1, Springer-Verlag, Berlin.

1961

Datta, B. K.: "Static Electromagnetic Fields in General Relativity," *Ann. Physics*, **12**, 295.

Finkelstein, R.: "Spinor Fields in Spaces with Torsion," *Ann. Physics*, **12**, 200.

Goldberg, J. N., and R. P. Kerr: "Some Applications of the Infinitesimal-Holonomy Group to the Petrov Classification of Einstein Spaces," *J. Mathematical Phys.*, **2**, 327.

Kalman, G.: "Lagrangian Formalism in Relativistic Dynamics," *Phys. Rev.*, **123**, 384.

Kerr, R. P., and J. N. Goldberg: "Einstein Spaces with Four-parameter Holonomy Groups," *J. Mathematical Phys.*, **2**, 332.

Klauder, J. R.: "Covariant Quantization of the Gravitational Field," *Il Nuovo Cimento*, **19**, 1059.

Newman, E.: "Some Properties of Empty Space-Time," *J. Mathematical Phys.*, **2**, 324.

Peres, A.: "On Geometrodynamics and Null Fields," *Ann. Physics*, **11**, 419.

Salam, A., and J. C. Ward: "On the Gauge Theory of Elementary Interactions," *Il Nuovo Cimento*, **19**, 1.

Sandage, A.: "The Ability of the 200-inch Telescope to Discriminate between Selected World Models," *Astrophys. J.*, **133**, 355.

Tangherlini, F. R.: "On the Energy-Momentum Tensor of the Gravitational Field," *Il Nuovo Cimento*, **20**, 1.

Toupin, R. A., and R. S. Rivlin: "Electro-magneto-optical Effects," *Arch. Rational Mech. Anal.*, **7**, 434.

INDEX

Index

Other RAND Books

COLUMBIA UNIVERSITY PRESS, NEW YORK, NEW YORK

Bergson, Abram, and Hans Heymann, Jr., *Soviet National Income and Product, 1940–48*, 1954
Galenson, Walter, *Labor Productivity in Soviet and American Industry*, 1955
Hoeffding, Oleg, *Soviet National Income and Product in 1928*, 1954

THE FREE PRESS, GLENCOE, ILLINOIS

Dinerstein, Herbert S., and Leon Goure, *Two Studies in Soviet Controls: Communism and the Russian Peasant; Moscow in Crisis*, 1955
Garthoff, Raymond L., *Soviet Military Doctrine*, 1953
Goldhamer, Herbert, and Andrew W. Marshall, *Psychosis and Civilization*, 1953
Leites, Nathan, *A Study of Bolshevism*, 1953
Leites, Nathan, and Elsa Bernaut, *Ritual of Liquidation: The Case of the Moscow Trials*, 1954
The RAND Corporation, *A Million Random Digits with 100,000 Normal Deviates*, 1955

HARVARD UNIVERSITY PRESS, CAMBRIDGE, MASSACHUSETTS

Bergson, Abram, *The Real National Income of Soviet Russia Since 1928*, 1961
Fainsod, Merle, *Smolensk under Soviet Rule*, 1958
Hitch, Charles J., and Roland McKean, *The Economics of Defense in the Nuclear Age*, 1960
Moorsteen, Richard, *Prices and Production of Machinery in the Soviet Union, 1928–1958*, 1962

THE MACMILLAN COMPANY, NEW YORK, NEW YORK

Dubyago, A. D., *The Determination of Orbits*, translated from the Russian by R. D. Burke, G. Gordon, L. N. Rowell, and F. T. Smith, 1961
O'Sullivan, J. J. (ed.), *Protective Construction in a Nuclear Age*, 1961
Whiting, Allen S., *China Crosses the Yalu: The Decision To Enter the Korean War*, 1960

McGRAW-HILL BOOK COMPANY, INC., NEW YORK, NEW YORK

Bellman, Richard, *Introduction to Matrix Analysis*, 1960
Dorfman, Robert, Paul A. Samuelson, and Robert M. Solow, *Linear Programming and Economic Analysis*, 1958

Gale, David, *The Theory of Linear Economic Models*, 1960
Janis, Irving L., *Air War and Emotional Stress: Psychological Studies of Bombing and Civilian Defense*, 1951
Leites, Nathan, *The Operational Code of the Politburo*, 1951
McKinsey, J. C. C., *Introduction to the Theory of Games*, 1952
Mead, Margaret, *Soviet Attitudes toward Authority: An Interdisciplinary Approach to Problems of Soviet Character*, 1951
Scitovsky, Tibor, Edward Shaw, and Lorie Tarshis, *Mobilizing Resources for War: The Economic Alternatives*, 1951
Selznick, Philip, *The Organizational Weapon: A Study of Bolshevik Strategy and Tactics*, 1952
Shanley, F. R., *Weight-Strength Analysis of Aircraft Structures*, 1952
Williams, J. D., *The Compleat Strategyst: Being a Primer on the Theory of Games of Strategy*, 1954

THE MICROCARD FOUNDATION, MADISON, WISCONSIN

Baker, C. L., and F. J. Gruenberger, *The First Six Million Prime Numbers*, 1959

NORTH-HOLLAND PUBLISHING COMPANY, AMSTERDAM, HOLLAND

Arrow, Kenneth J., and Marvin Hoffenberg, *A Time Series Analysis of Interindustry Demands*, 1959

FREDERICK A. PRAEGER INC., NEW YORK, NEW YORK

Dinerstein, H. S., *War and the Soviet Union: Nuclear Weapons and the Revolution in Soviet Military and Political Thinking*, 1959
Speier, Hans, *Divided Berlin: The Anatomy of Soviet Political Blackmail*, 1961
Tanham, G. K., *Communist Revolutionary Warfare: The Viet Minh in Indochina*, 1961

PRENTICE-HALL, INC., ENGLEWOOD CLIFFS, NEW JERSEY

Dresher, Melvin, *Games of Strategy: Theory and Applications*, 1961
Newell, Allen (ed.), *Information Processing Language-V Manual*, 1961
Hsieh, Alice L., *Communist China's Strategy in the Nuclear Era*, 1962

PRINCETON UNIVERSITY PRESS, PRINCETON, NEW JERSEY

Baum, Warren C., *The French Economy and the State*, 1958
Bellman, Richard, *Adaptive Control Processes: A Guided Tour*, 1961
Bellman, Richard, *Dynamic Programming*, 1957
Bellman, Richard E., and Stuart E. Dreyfus, *Applied Dynamic Programming*, 1962
Brodie, Bernard, *Strategy in the Missile Age*, 1959
Davison, W. Phillips, *The Berlin Blockade: A Study in Cold War Politics*, 1958
Ford, L. R., Jr., and D. R. Fulkerson, *Flows in Networks*, 1962
Hastings, Cecil, Jr., *Approximations for Digital Computers*, 1955
Johnson, John J. (ed.), *The Role of the Military in Underdeveloped Countries*, 1962
Smith, Bruce Lannes, and Chitra M. Smith, *International Communication and Political Opinion: A Guide to the Literature*, 1956
Wolf, Charles, Jr., *Foreign Aid: Theory and Practice in Southern Asia*, 1960

PUBLIC AFFAIRS PRESS, WASHINGTON, D.C.

Krieger, F. J., *Behind the Sputniks: A Survey of Soviet Space Science*, 1958
Rush, Myron, *The Rise of Khrushchev*, 1958

RANDOM HOUSE, INC., NEW YORK, NEW YORK

Buchheim, Robert W., and the Staff of The RAND Corporation, *Space Handbook: Astronautics and Its Applications*, 1959

ROW, PETERSON AND COMPANY, EVANSTON, ILLINOIS

George, Alexander L., *Propaganda Analysis: A Study of Inferences Made from Nazi Propaganda in World War II*, 1959
Melnik, Constantin, and Nathan Leites, *The House without Windows: France Selects a President*, 1958
Speier, Hans, *German Rearmament and Atomic War: The Views of German Military and Political Leaders*, 1957
Speier, Hans, and W. Phillips Davison (eds.), *West German Leadership and Foreign Policy*, 1957

STANFORD UNIVERSITY PRESS, STANFORD, CALIFORNIA

Goure, Leon, *The Siege of Leningrad, 1941–1943*, 1962
Kecskemeti, Paul, *Strategic Surrender: The Politics of Victory and Defeat*, 1958
Kecskemeti, Paul, *The Unexpected Revolution: Social Forces in the Hungarian Uprising*, 1961
Kramish, Arnold, *Atomic Energy in the Soviet Union*, 1959
Leites, Nathan, *On the Game of Politics in France*, 1959
Trager, Frank N. (ed.), *Marxism in Southeast Asia: A Study of Four Countries*, 1959

UNIVERSITY OF CALIFORNIA PRESS, BERKELEY, CALIFORNIA

Goure, Leon, *Civil Defense in the Soviet Union*, 1962

THE UNIVERSITY OF CHICAGO PRESS, CHICAGO, ILLINOIS

Hirshleifer, Jack, James C. DeHaven, and Jerome W. Milliman, *Water Supply: Economics, Technology, and Policy*, 1960

JOHN WILEY & SONS, INC., NEW YORK, NEW YORK

McKean, Roland N., *Efficiency in Government through Systems Analysis: With Emphasis on Water Resource Development*, 1958